F. Preparata (Ed.)

Theoretical Computer Science

Lectures given at a Summer School of the
Centro Internazionale Matematico Estivo (C.I.M.E.),
held in Bressanone (Bolzano), Italy,
June 9-17, 1975

FONDAZIONE
CIME
ROBERTO CONTI

 Springer

C.I.M.E. Foundation
c/o Dipartimento di Matematica "U. Dini"
Viale Morgagni n. 67/a
50134 Firenze
Italy
cime@math.unifi.it

ISBN 978-3-642-11118-1   e-ISBN: 978-3-642-11120-4
DOI:10.1007/978-3-642-11120-4
Springer Heidelberg Dordrecht London New York

Printed on acid-free paper

Springer.com

CENTRO INTERNAZIONALE MATEMATICO ESTIVO

(C.I.M.E.)

I Ciclo - Bressanone dal 9 al 17 giugno 1975

THEORETICAL COMPUTER SCIENZE

Coordinatore: Prof. F. PREPARATA

CENTRO INTERNAZIONALE MATEMATICO ESTIVO
(C.I.M.E.)

PARALLEL PROGRAM SCHEMATA

R.E. MILLER

Corso tenuto a Bressanone dal 9 al 17 giugno 1975

R. E. Miller

Configurable Computers and the Data Flow
Model Transformation

R. E. Miller

In this lecture we discuss a type of computer organization which is based upon the concept of operation sequencing being controlled by operand availability. Such sequencing differs radically from current computer organizations in which the sequencing is determined by an explicit or implicit ordering of the instructions which is controlled through an instruction counter which specifies the next instruction to be performed. We will discuss a particular type of data sequenced computer called "configurable computers" [7]. Closely related types of machines have subsequently also been proposed and studied by others [3, 11, 10, 9].

One of the major problems with unconventional computer organizations that have been proposed in the past is the great difficulty of programming such machines. In some cases the proposed structures were so complex that an inordinate amount of work was required to specify what each object in the structure was to do, while in other cases the structures were particularly suited to only a narrow class of computations, and were ill suited for general purpose use. A number of approaches for representing the computational process in a data sequenced manner have recently been proposed [11, 5, 2, 1]. We will briefly describe one such approach [8, 6] which is aimed at automatically transforming normal computer programs into a data sequenced form suitable for configurable computers. Even though a direct representation of an algorithm into a data sequenced form may be able to better exploit the advantages of the data sequenced form, consider-able advantage is gained in automatic transformation -- closely related to optimizing compiler techniques -- which allows the user to continue to use his well known programming languages directly, rather than learning a new complex language immediately.

R. E. Miller

Configurable Computers

One of the central concepts of configurable computers is to have a machine that changes its structure into the natural structure of the algorithm being performed, thereby allowing parallel computations and many of the speed advantages of special purpose devices.

Two particular approaches for attaining this goal are described. The one approach is called Search Mode and the other is called Interconnection Mode. These should be viewed as only two ends of a spectrum of possibilities in which considerable differences in the possible computer control structures exist.

Search Mode Configurables

The basic organization of a search mode machine is depicted in Figure 1.

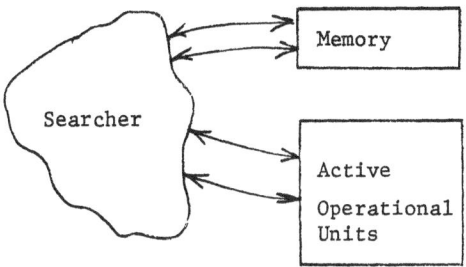

The operational units are thought of as a set of either general or special purpose units which perform the computational, conditional, and data generation aspects of the computation. When one of these units has completed a task it requests the searcher to find another task for it to perform. The searcher, which is essentially a new kind of control, then inspects memory, or a suitable portion thereof, for a new task for the operational unit. This organization can thus be adopted to the idea that a task becomes capable of being performed when its operands have

been computed. An example of machine instruction format aids in seeing in
more detail how the machine could operate. We give an example for an arithmetic operation but, of course, a complete repetoire of instructions and
formats would have to be given for a complete specification.

Operation Code	Status Bits	First Operand	Second Operand	"Address" for Result

This instruction format contains first a field of bits to specify the
type of operation to be performed (the operation code). Skipping over the
status bits for a moment, it then contains fields for the operand values
to be stored. Within the status bits one keeps track of whether the
operands currently reside in these locations (i.e., have been computed and
stored there) and whether the location for the result is available for
storing a result. Finally, the "address for result" field specifies where
the result is to be stored (i.e., as an operand for some other instruction)
Storing a result in an operand field updates the status bits, as does the
action of performing an instruction. Clearly, this type of format
eliminates the need for normal instruction sequencing. The sequencing is
essentially "data driven."

It is the operation code and status bit fields that the searcher
inspects to determine readiness of an instruction. This could be imagined
to be done by a type of associative search, or by a method of building up
stacks or queues of instructions that are ready to be performed. Before
describing how programs can be transformed into such a data sequenced
form, we describe the interconnection mode machine idea.

R. E. Miller

Interconnection Mode Configurables

In the search mode concept the sequencing of instructions was done through the action of operation results becoming operands of new instructions, and these transfers of information were done by storing results in the proper places of instructions. With the advent of economical electronic switches more direct connection of result to operand could be envisioned. This is the case for the interconnection mode idea. A block diagram is shown below.

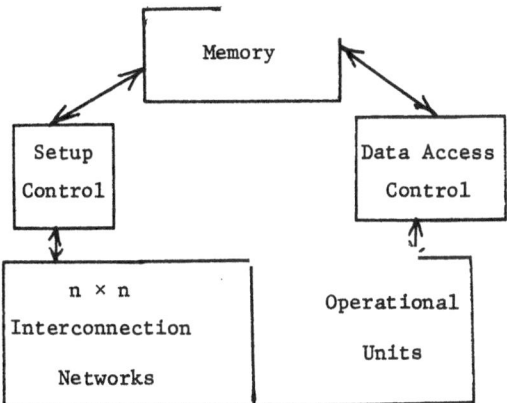

Here the interconnection network is used to directly interconnect outputs of one operational unit to inputs of another operational unit in accord with the result to operand specifications of the algorithm. The basic steps for such a machine are:

1. Decompose the program into appropriate size blocks.

2. Transform each block into a data sequenced form.

3. Store the blocks, so transformed, in memory as setup instructions for the interconnection network.

4. Choose a block to be performed (to start with this is the block with the start of the program –– subsequently this is specified

by what block the running block exits to next) and set up the
interconnections as specified.

5. Perform the block execution. Note that during this time no
    instuctions -- only data -- need be stored or fetched from memory.
6. Termination of block specifies next block to be performed (return
    to 4).

At this point it should be clear that both the search mode and interconnection mode machines are sequenced essentially by data paths rather
than control paths through the algorithm. Thus, the natural parallelism
of the algorithm can be used to speed up operation. Also, the machines
have some of the speed advantages of special purpose devices. This is
especially true of the interconnection mode machine since units are actually directly interconnected as they would be in a special purpose device
The operational control -- distributed throughout the machine by data
availability considerations -- is also rather simple, and this could be a
distinct advantage over other approaches to high performance where very
complex control mechanisms are required. Finally, as we have indicated,
standard programming languages can be used to express the jobs to be done
and transformational techniques to provide suitable machine language
instructions for these machines should be readily developed using techniques known for compiler optimization. Other potential advantages for
these machines are given in [7].

Data Sequenced Program Transformation

We now outline how a program can be transformed into a form suitable
for our configurable computers. We call this a data flow transformation.
The basic steps of the transformation are:

1. Partition the program into "basic blocks" and name each block. A
    basic block is a contiguous segment of code which can be entered only
    through the first instruction of the block, must be executed by execut-
    ing each successive instruction in order, and can be exited only from
    the last instruction in the block. I.e., it is a "straight line"
    segment of code.

2. Determine the immediate predecessors and successors of each basic
    block.

3. Generate (in arbitrary order) a "data flow segment" for each block.
    A data flow segment consists of:

    (i)   input list -- i.e. variables needed by the block.

    (ii) output list -- i.e. result names at end of block execution.

    (iii) interconnected modules -- i.e. the operations and flow of data
          from result to operand between operations in the block.

4. Interconnect data flow segments. This uses the predecessor and
    successor information and updates input and output lists for data
    "passing through" a block.

    Without going into great detail we illustrate the transformation
through an example from [6].

An Example of Data Flow Model Transformation

As an example program we consider the problem of evaluating the function $f(x) = a^x + bx + c$. We assume that x, a, b, and c are inputs
stored in the symbolic locations x, a, b, and c respectively, and also
assume that x is a positive integer. A simple program to perform this
evaluation is shown below. The program language used is simple and should
be self-explanatory.

Statement #	Program	Comments
1	CLA x	set accumulator to x.
2	STO COUNT	put x in location COUNT.
3	CLA a	put a in accumulator.
4	TRA 6	transfer to statement 6.
5	MPY a	multiply accumulator by a.
6	Decrement COUNT	decrease COUNT by 1.
7	Branch on COUNT (to 5 on $\neq$ 0)	conditional transfer.
8	STO T	store a^x in T.
9	CLA x	place x in accumulator.
10	MPY b	form bx in accumulator.
11	ADD T	form a^x+bx in accumulator.
12	ADD C	form a^x+bx+c in accumulator.
13	STO R	store result in R.

Applying the notion of basic block to this program we find that instructions 1, 2, 3, 4 form a basic block with instruction 1 being the start of the program. Similarly instructions 6, 7 form a basic block, 8, 9, 10, 11, 12, 13 form a basic block and 5 alone forms a basic block. These blocks are depicted and named BB1 through BB4 in the following diagram.

R. E. Miller

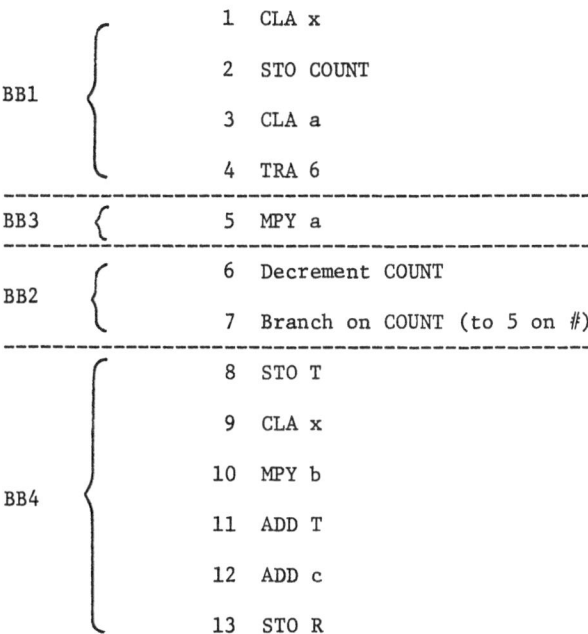

BB1	1	CLA	x
	2	STO	COUNT
	3	CLA	a
	4	TRA	6
BB3	5	MPY	a
BB2	6	Decrement COUNT	
	7	Branch on COUNT (to 5 on #)	
BB4	8	STO	T
	9	CLA	x
	10	MPY	b
	11	ADD	T
	12	ADD	c
	13	STO	R

Step 2 of the algorithm determines immediate successors and immediate predecessors as shown in the following figure.

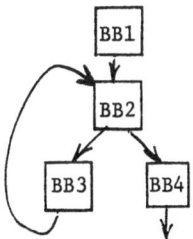

Step 3 of the algorithm generates a "data flow segment" for each basic block. The idea here is to generate a list of items needed as inputs to the block, the outputs created by the block and the operations used to create these outputs along with the flow of data between the operations within the block. The items have names associated with them through the program language definition, these names we call "source names." During generation we assign "local data names" to items also. Consider, for

example, basic block 2, (BB2). This block starts with instruction 6 —
Decrement COUNT. By definition this instruction needs an input with source
name "COUNT" and produces a new output also called "COUNT" which has a
value less than the original value of COUNT. Thus COUNT is placed on the
input list, and we associate a local data name with this input value. We
use the name BB2,1; i.e., the first local data name in BB2. In the output list we then have COUNT with a new value and name this new value BB2,2
as a local data name. The operation performed is a SUBTRACT 1 so this
operation gets placed in the module structure with input BB2,1 and output
BB2,2. The second instruction of BB2 is Branch on COUNT. This instruction uses the current value of COUNT, namely BB2,2 and tests it for =0
or ≠0. This is indicated in the output list as changing item COUNT-BB2,2
to two values COUNT BB2,2T1 and COUNT BB2,2T2 for the outcome of the
test either being outcome T1 or T2. A test module is added to the
module structure -- we call it test T -- with the two indicated outputs.
This completes Step 3 for BB2. Our result is summerized below.

BB2   Data Flow Segment

Input List   COUNT - BB2,1

Output List   ~~COUNT = BB2,2~~

COUNT - BB2,2T1

COUNT - BB2,2T2

Module Structure

Similar calculations are done for each of the other basic blocks producing the following results.

BB1 Data Flow Segment

Input List	x - BB1,1
	a - BB1,2
Output List	~~ACCUM = BB1,1~~
	COUNT - BB1,1
	ACCUM - BB1,2

no modules.

BB3 Data Flow Segment

Input List	ACCUM - BB3,1
	a - BB3,2
Output List	ACCUM - BB3,3

Module
Structure

BB3,1   BB3,2

MPY

BB3,3

BB4 Data Flow Segment

Input List	ACCUM - BB4,1
	x - BB4,2
	b - BB4,3
	c - BB4,6
Output List	T - BB4,1
	~~ACCUM = BB4,2~~
	~~ACCUM = BB4,4~~
	~~ACCUM = BB4,5~~
	ACCUM - BB4,7
	R - BB4,7

Module
Structure

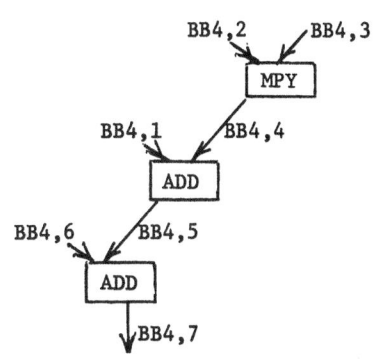

Step 4 of the transformation interconnects these module structures by using
successor and predecessor information and making output to input connections
through common source names. The result of making these interconnections
and inserting test points T1 and T2 for places where data passes only
conditionally on the outcome of test T is shown in the next figure. Note
that even in this simple example some possibilities for parallelism are
exhibited. For example, the two multiplications and the subtract 1
operations could all be performed concurrently.

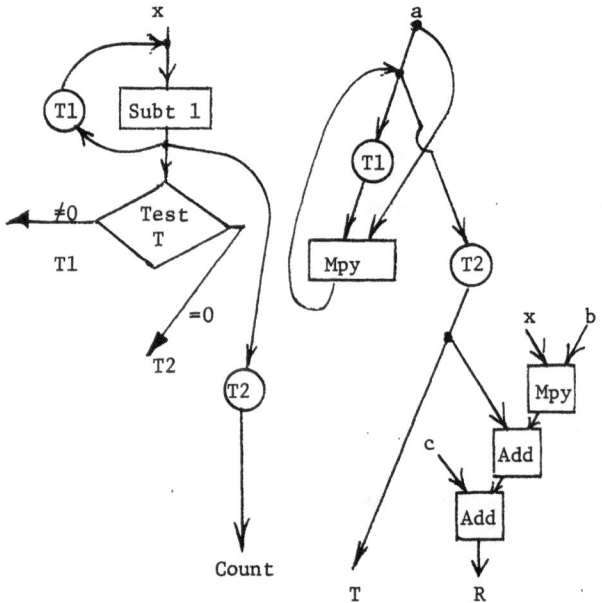

It should be clear from the descriptions given for configurable
computers that this diagram provides all the essential information needed
to specify the instructions for a search mode machine or the interconnections for an interconnection mode machine. Clearly, the sequencing so
specified differs considerably from how the original program would have
run on a conventional computer.

R. E. Miller

REFERENCES

[1] A. Bährs, "Operation Patterns," Symposium on Theoretical Programming, Novosibirsk, USSR, August 1972, in Lecture Notes in Computer Science, Vol. 5 International Symposium on Theoretical Programming, Springer-Verlag, New York, 1974, pp. 217-246.

[2] J. B. Dennis, J. B. Fosseen, and J. P. Linderman, "Data Flow Schemas,' Symposium on Theoretical Programming, Novosibirsk, USSR, August 1972, in Lecture Notes in Computer Science, Vol. 5 International Symposium on Theoretical Programming, Springer-Verlag, New York, 1974, pp. 187-216.

[3] J. B. Dennis and D. P. Misunas, "A Computer Architecture for Highly Parallel Signal Processing," Proceedings ACM Annual Conference, November 1974, pp. 402-409.

[4] K. B. Irani and C. R. Sonnenburg, "Exploitation of Implicit Parallelism in Arithmetic Expressions for an Asynchronous Environment," Report

[5] P. R. Kosinski, "A Data Flow Programming Language," IBM T. J. Watson Research Center Report RC-4264, Yorktown Heights, N. Y., March 1973.

[6] R. E. Miller, "The Data Flow Model Transformation," to appear in the Proceedings of the 1972 GMD Summer Seminar on Systems Organization, Bonn, Germany.

[7] R. E. Miller and J. Cocke, "Configurable Computers: A New Class of General Purpose Machines," Symposium on Theoretical Programming, Novosibirsk, USSR, August 1972, in Lecture Notes in Computer Science, Vol. 5 International Symposium on Theoretical Programming, Springer-Verlag, New York, 1974, pp. 285-298.

[8] R. E. Miller and J. D. Rutledge, "Generating a Data Flow Model of a Program," IBM Technical Disclosure Bulletin, Vol. 8, No. 11, April 1966, pp. 1550-1553.

[9] S. S. Reddi and E. A. Feustel, "A Restructurable Computer System," Report, Laboratory for Computer Science and Engineering, Rice Univ., Houston, Texas, March 1975.

[10] C. R. Sonnenburg, "A Configurable Parallel Computing System," Ph.D. Dissertation, University of Michigan, Ann Arbor, October 1974.

[11] J. C. Syre, "From the Single Assignment Software Concept to a New Class of Multiprocessor Architectures," Report, 1975 Department d'Informatique, C.E.R.T. BP4025, 31055 Toulouse Cedex, France.

R. E. Miller

Computation Graphs and Petri Nets

R. E. Miller

In this lecture we show how a special type of Petri net, widely studied in the literature, can be represented by computation graphs. We then illustrate how results about computation graphs can be translated into results for such Petri nets. We first introduce the two models.

Computation Graphs [3]

A computation graph G is a finite directed graph consisting of: (i) nodes $n_1, n_2, \ldots, n_\ell$; (ii) edges $d_1, d_2, \ldots, d_t$, where any given edge d_p is directed from a specified node n_i to a specified node n_j; (iii) four non-negative integers A_p, U_p, W_p and T_p, where $T_p \geq W_p$, associated with each edge d_p.

Nodes represent operations, edges represent first-in-first-out queues of data, and for an edge d_p directed from n_i to n_j the four parameters mean: A_p is the initial number of items in the queue, U_p is the number of items added to the queue each time n_i fires, W_p is the number of items removed from the queue each time n_j fires, and T_p is the number of items required as operands for n_j to fire.

The idea of computation sequences for a computation graph is that an operation can fire whenever it has a sufficient number of operands on each of its incoming edges. After firing it places results on each of its outgoing edges, and these results may be used later as operands for other operations. Some simple examples of computation graphs are shown in the following figures:

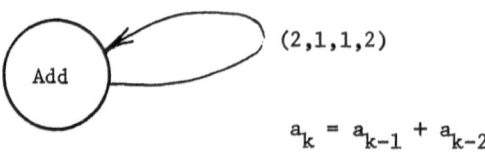

$$a_k = a_{k-1} + a_{k-2}$$

R. E. Miller

This one node, one edge example shows how a computation graph can realize the computation of successive Fibonacci numbers when the two initial values are each 1. Note here that although $T = 2$ here $W = 1$ so that two operands are needed for the add operation but only one is removed. Thus the second operand in one firing becomes the first operand in the next firing.

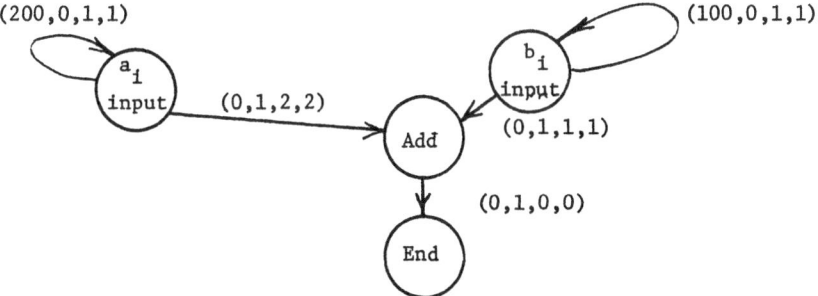

This example illustrates combining two lists of numbers A and B to form a list C according to the equation

$$c_i = a_{2i-1} + a_{2i} + b_i \quad \text{for } i = 1,2,\ldots,100.$$

More complex computation graphs are shown in [3 & 5], where they are studied in more detail.

Petri Nets [6]

Petri nets have become a very popular means of representing parallelism in systems. Some examples of the literature are [1, 2, 6, 7]. It is a simple directed graph model in which the notion of computation can be easily explained. More formally, a Petri net $P = (\Pi, \Sigma, R, M_0)$ consists of a finite set of nodes Π of places, a finite set of nodes Σ of transitions, a relation $R \subseteq (\Pi \times \Sigma) \cup (\Sigma \times \Pi)$ which indicates directed edges between nodes, and a mapping M_0 from Π to the set of non-negative integers, called the <u>initial marking</u> which places a number of "tokens" in each place of P. (See [5] for a more complete definition.) The tokens in places are used to control the firing of transitions of a Petri net, and it is these firings that represent computations. The set of places having edges directed into a given transition are called the <u>input places</u> for the transition, and the set of places having edges directed out of the transition are called its <u>output places</u>. A transition is <u>active</u> if each of its input places contains one or more tokens. An active transition may <u>fire</u>. This removes one token from each of its input places and adds one token to each of its output places. A token in a given place can be used in only one transition firing. A sequence of transition firings is called a <u>firing sequence</u> and each firing changes the <u>marking</u> (tokens) in the places.

A simple Petri net is shown in the next figure. Here circles represent places, dots in the circles - tokens, and bars represent transitions.

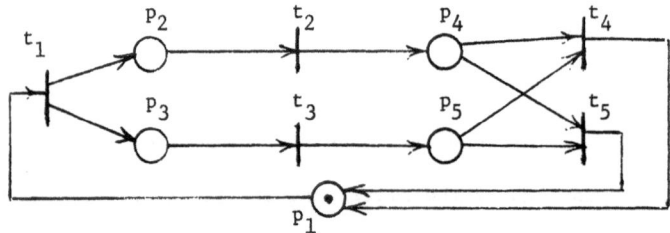

As shown t_1 is active. When it fires it removes the token from p_1 and places tokens in p_2 and p_3 Now t_2 and t_3 are active. After they both fire tokens reside in p_4 and p_5 only. Now both t_4 and t_5 are active, but only one can fire since there is only one token each in p_4 and p_5. This is called a <u>conflict</u>. In such a case an arbitrary choice is made, and the firing places a token back in p_1, as was the case at the start.

R. E. Miller

If we represent firing sequences by sequences of transition letters, then the reader may check that the following are firing sequences for this example:

$$t_1 t_2 t_3 t_4 t_1 \cdots$$

$$t_1 t_3 t_2 t_4 \cdots$$

$$t_1 t_3 t_2 t_5 \cdots$$

$$t_1 t_2 t_3 t_5 \cdots$$

Many special types of Petri nets have been studied. One such type restricts each place to have exactly one input transition and one output transition. These kind of Petri nets are called <u>marked graphs</u>. A simple example is shown below.

One can now imagine removing the places from marked graphs and letting tokens reside on edges. Then the figure simplifies to:

which is a directed graph with only one type of node.

Marked Graphs and Computation Graphs

It is easy to see [4] that any marked graph can be represented by a computation graph with

$$U_p = W_p = T_p = 1$$

for each edge, and A_p determined by the initial marking.

Our example marked graph reduces to the following computation graph:

R. E. Miller

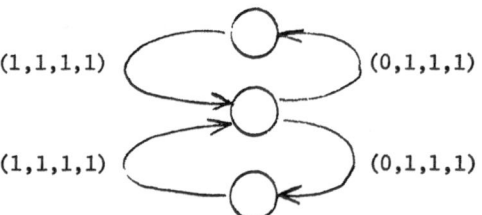

Through this reduction properties of computation graphs are readily translated to properties of marked graphs. For example, for any loop $d_1, d_2 \ldots d_\ell$ of a computation graph with $\dfrac{U_1 \ldots U_\ell}{W_1 \ldots W_\ell} = 1$, it is known that

$$q(1,N) + \frac{W_1}{U_2} \; q(2,N) + \ldots + \frac{W_1 \ldots W_{\ell-1}}{U_2 \ldots U_\ell} \; q(\ell,N) \text{ is constant, where } q(i,N) \text{ is}$$

the queue length on edge d_i. This translates to say that: The token count around a loop in a marked graph does not change by transition firing.

We thus have seen, through a simple relationship between models, how a result can be translated to the other model. We are now ready to consider other models.

References

[1] F. Commoner, A. W. Holt, S. Even, A. Pneuli, "Marked Directed Graphs, Journal of Computer and Systems Sciences, 5, October 1971, pp.511-23.

[2] M. Hack, "Decision Problems for Petri Nets and Vector Addition System," MIT Project MAC, Computation Structures Group Memo 95, March, 1974.

[3] R.M. Karp and R.E. Miller, "Properties of a Model for Parallel Computation: Determinacy, Termination, Queueing," SIAM Journal of Applied Math., 1390-1411, November, 1966.

[4] R.E. Miller, "A Comparison of Some Theoretical Models of Parallel Computation," IEEE Trans. on Computers, C-22, 8, 710-717, Aug. 1973.

[5] R.E. Miller, "Some Relationships Between Various Models of Parallelism and Synchronization," IBM Research Report, RC 5074, October 1974.

[6] C.A. Petri, "Communication with Automata," Supplement 1 to Technical Report RADC-TR-65-337, 1, Griffiss Air Force Base, New York, 1966 [Translated from German "Kommunikation mit Automaten," Univ. of Bonn, 1962]

R. E. Miller

[7] M. Yoeli, "Petri Nets and Asynchronous Control Networks," Applied
     Analysis and Computer Science Research Report CS-73-07, University
     of Waterloo, Waterloo, Ontario, Canada, April, 1973.

R. E. Miller

Parallel Program Schemata

R. E.Miller

Introduction

The parallel program schema is an abstract formulation of parallel programs. It is a complex model which incorporates some very general notions
of operation sequencing over a finite set of operations which fetch and
store data in a random access memory. It is an abstraction (thus called a
schema) in the sense that it leaves unspecified certain of the functional
aspects of the operations and concentrates on the sequencing aspects of
the computations and those properties that hold true no matter how the
functional aspects are defined. In this series of five lectures we introduce the model, illustrate some of the properties that can be studied and
prove some results indicating the techniques used to prove decidability
and undecidability of the properties. See [1-5] for more detail.

Definitions and Basic Properties

<u>Definition</u>   A parallel program schema $\mathcal{S}$ = (M, A, $\mathcal{J}$) consists of

   (i)   M, a set of <u>memory</u> <u>locations</u>

   (ii)   A, a finite set of operations A = {a, b, c,$\cdots$ }

         where for each   a$\in$A   we have:

         (a)   a positive integer K(a) called the number of <u>outcomes</u>,

         (b)   D(a) $\subseteq$ M a specified set of <u>domain</u> <u>locations</u>,

         (c)   R(a) $\subseteq$ M a specified set of <u>range</u> <u>locations</u>.

(iii) $\mathcal{J}$ = (Q, q_0, Σ, τ), the <u>control</u>, where:

      Q is a set of <u>states</u>

      $q_0 \in Q$ is the <u>initial</u> <u>state</u>

      $\Sigma = \Sigma_i \cup \Sigma_t$, the <u>event</u> <u>alphabet</u>

            with $\Sigma_i = \bigcup_{a \in A} \{\bar{a}\}$, the <u>initiation</u> <u>symbols</u>

            and $\Sigma_t = \bigcup_{a \in A} \{a_1, a_2, \cdots, a_{K(a)}\}$, the <u>termination</u> <u>symbols</u>.

τ is the <u>transition function</u> which is a partial function from

$Q \times \Sigma$ into Q and which is total on $Q \times \Sigma_t$.

A schema is thought to operate in the following manner. A computation
is a sequence of events -- i.e., initiations and terminations of operations.
Upon initiation of an operation a it reads its operand values from locations
$D(a)$. Sometime after initiating the performance of the operation terminates.
This is indicated by some termination symbol $a_j \epsilon \Sigma_t$. Upon termination the
results of operation a are stored into the $R(a)$ locations. Also the
symbol a_j indicates that outcome   j   (or conditional branch outcome   j)
occurs. Formally termination a_j creates a change of state in the control
$\mathcal{J}$ and from this new state other initiations may be defined. To define
these ideas more completely we give several further definitions.

<u>Definition:</u>   An <u>interpretation</u> $\mathcal{A}$ of $\mathcal{S}$ consists of:

(i)   a function   C   associating a set $C(i)$ with each $i \epsilon M$,

        specifying the set of values allowed,

(ii)   initial memory contexts $c_0 \epsilon X_{i \epsilon M} C(i)$.

(iii) For each $a \epsilon A$, two functions:

$$F_a : \underset{i \epsilon D(a)}{X} C(i) \longrightarrow \underset{i \epsilon R(a)}{X} C(i)$$

(This is the function that given values from $D(a)$ will

determine values for $R(a)$ locations.)

$$G_a : \underset{i \epsilon D(a)}{X} C(i) \longrightarrow \{a_1, a_2, \cdots, a_{K(a)}\}$$

(This function determines the outcome of the operation a

performance given values from $D(a)$.)

To define computations as sequences of event symbols from Σ we
let $\alpha = (c, q, \mu)$ designate an $\mathcal{A}$-instantaneous description of the
process. Here   c   is the current contents of memory; q   is the current

state of the control; and μ is a list for each $a \epsilon A$ where each item in
the list $\mu(a)$ consists of a tuple $X_{i \epsilon D(a)} C(i)$ of domain values of performances of a currently in process. The initial instantaneous description $\alpha_0 = (c_0, q_0, \mu_0)$ consists of the initial memory contents, the
initial state q_0, and the μ_0 lists are empty for each operation. The
basic sequencing of operations then is defined as transitions from one
$\mathcal{A}$-instantaneous description to another as follows:

<u>Definition</u>: The partial function $\alpha \cdot \sigma$, for α an $\mathcal{A}$-instantaneous
description, and $\sigma \epsilon \Sigma$ is defined by:

(1) (Initiation symbol case) $\sigma = \bar{a}$, $\alpha = (c, q, \mu)$:

   $\alpha \cdot \bar{a}$ is defined iff $\tau(q, \bar{a})$ is defined.

If so: $\alpha \cdot \bar{a} = (c', q', \mu')$ where

   $c' = c$, $q' = \tau(q, \bar{a})$ and

   for $b \neq a$ $\mu'(b) = \mu(b)$ and $\mu'(a)$ is the list $\mu(a)$

   with $X_{i \epsilon D(a)} c(i)$ added to the end of the list.

(2) (Termination symbol case) $\sigma = a_j$, $\alpha = (c, q, \mu)$:

   $\alpha \cdot a_j$ is defined iff $\mu(a)$ is nonempty and $G_a(\xi) = a_j$,

   where ξ denotes the first element in the $\mu(a)$ list.

   In this case $\alpha \cdot a_j = (c', q', \mu')$ where:

   (i)   for $i \notin R(a)$ $c'(i) = c(i)$.

   (ii) for $i \epsilon R(a)$ $c'(i)$ is the component of $F_a(\xi)$

         'corresponding to i.

   (iii) $q' = \tau(q, a_j)$

   (iv) $\mu'(b) = \mu(b)$ for $b \neq a$ and $\mu(a) = \xi \mu'(a)$;

         that is, $\mu'(a)$ is the remainder of $\mu(a)$ after

         the first element of $\mu(a)$ is deleted.

For $y \epsilon \Sigma^*$, $\alpha \cdot y$ is defined in the normal way by letting $\alpha \cdot x \sigma = (\alpha \cdot x) \cdot \sigma$.
This now lets us define schema $\mathcal{A}$-computations.

R. E. Miller

<u>Definition</u>: A finite or infinite word x over Σ is an $\mathcal{L}$-<u>computation</u>
iff:

(i)   for every prefix y of x $\alpha_0 \cdot y$ is defined;

(ii) if x is finite, then for all $\sigma \epsilon \Sigma$, $\alpha_0 \cdot x\sigma$ is undefined;

(iii) (Finite delay property) If y is a prefix of x and $\sigma \epsilon \Sigma$ with
the property that for every z such that yz is a prefix of x
$\alpha_0 \cdot yz\sigma$ is defined, then for some z' $yz'\sigma$ is a prefix of x.

Thus an $\mathcal{L}$-computation by (i) must be consistent with the $\alpha \cdot x$ definition (follow the transition rules), by (ii) can end only if no other event
is possible, and by (iii) says that if after some point an event can
"constantly" occur, then it eventually does occur after some finite delay.

If x is an $\mathcal{L}$-computation then

$$\psi(x) = \alpha_0, \; \alpha_0 \cdot_1 x, \; \alpha_0 \cdot_2 x, \cdots, \alpha_0 \cdot_k x, \cdots$$

is called the <u>history</u> of x.

(Here $_i x$ is the prefix of length i of x.) $\psi_i(x)$ denotes the subsequence of $\psi(x)$ containing α_0 and the successive values of $\psi(x)$
that "store values" in location i. For $\alpha = (c, q, \mu)$ we let
$\Pi_i(\alpha) = c(i)$, $i \epsilon M$, $\Pi_a(\alpha) = \mu(a)$, $a \epsilon A$, etc. We denote by $\Omega_i(x) = \Pi_i(\psi_i(x))$
and call this the <u>contents sequence of cell i for x</u>. $\Omega_i(x)$ gives the
successive values that appear in location i during computation x.

<u>Definition</u>: A schema $\mathcal{S}$ is called <u>determinate</u> if whenever x and y
are $\mathcal{L}$-computations for some interpretation $\mathcal{L}$, then:

$$\forall i \epsilon M [\Omega_i(x) = \Omega_i(y)]$$

This is one of the basic properties of schemata. It is a property
that indicates a kind of "proper behavior." In references [1,2] this
property and others are studied in detail. For our lectures we concentrate
mainly on the determinacy property.

R. E. Miller

Before we proceed to prove some results about determinacy, a simple example schema might be helpful.

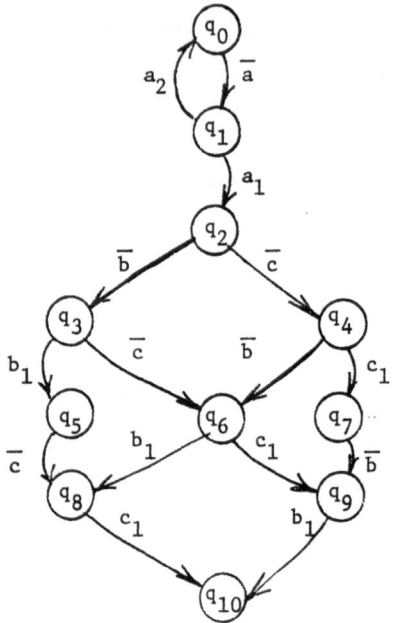

This example shows the control aspects of a schema with three operations $\{a, b, c\}$. All initiation symbol transitions are shown, but only those termination symbol transitions that can actually occur in computations are shown. Here operation a has two outcomes a_1 and a_2, and the other operations have only one outcome. This depicts the following situation. Operation a starts the computation. As long as it has outcome a_2 only a sequence of a operations can be performed. When outcome a_1 occurs we exit from the "a loop" and allow b and c to operate in parallel. Essentially we FORK at this point, allowing b and c to be performed and then on termination of both b and c performances we JOIN (in state q_{10}). Some possible computations are:

(1) $\overline{a}\ a_2\ \overline{a}\ a_1\ \overline{b}\ b_1\ \overline{c}\ c_1$

(2) $\overline{a}\ a_1\ \overline{b}\ \overline{c}\ c_1\ b_1$

(3) $\overline{a}\ a_1\ \overline{c}\ \overline{b}\ c_1\ b_1$

In computation (1) no parallelism occurs, that is only one operation is in execution at any moment. In computations (2) and (3) however, operations b and c are, after the fourth symbol of the sequence, concurrently being performed. The particular computations which actually would occur depend upon the domain and range location specifications for each operation, the F and G specifications in the interpretation as well as other aspects of the interpretation.

Necessary and Sufficient Conditions for Determinacy

We have just defined a property of schemata called determinacy. Essentially this means that no matter what the interpretation, any computations for that interpretation will provide the same sequence of values to occur in any memory cell. This, of course, implies that for terminating computations the final results are independent of the sequencing, but is even stronger than that, insisting that the sequence of intermediate values is also. We now wish to prove the following theorem about determinacy.

Theorem: Let $\mathcal{S}$ be a persistent, commutative, lossless schema. Then $\mathcal{S}$ is determinate if and only if for all interpretations $\mathcal{I}$ condition (A) holds:

(A) If $\alpha_0 \cdot u\sigma\pi$ and $\alpha_0 \cdot u\pi\sigma$ are both defined, then

$$\alpha_0 \cdot u\sigma\pi = \alpha_0 \cdot u\pi\sigma,$$

where $\pi,\ \sigma\epsilon\Sigma,\ u\epsilon\Sigma^*$.

What does this theorem say intuitively?

It says that for a certain type of schema -- conditions we will define -- if we reach a point at which two events can occur "simultaneously" or in either order then the result of this race is not dependent upon the winner of the race.

The special conditions of the hypothesis are the following.

Definition: A schema is <u>persistent</u> if and only if whenever σ and π are distinct elements of Σ and $\tau(q,\sigma)$ and $\tau(q,\pi)$ and both defined, then $\tau(q,\sigma\pi)$ and $\tau(q,\pi\sigma)$ are also defined.

Definition: A schema $\mathcal{S}$ is <u>commutative</u> if and only if whenever $\tau(q,\sigma\pi)$ and $\tau(q,\pi\sigma)$ are both defined then they are equal.

Definition: A schema $\mathcal{S}$ is <u>lossless</u> if for all $a \in A$, $R(a) \neq \emptyset$.

The proof of the theorem requires a number of definitions and lemmas. The first step is to show that we only need to consider a subclass of all interpretations, called one-one interpretations, to prove the theorem. Essentially, a one-one interpretation is an interpretation that stores in each cell the complete history of events that effect the cell. The needed lemma is then:

Lemma 1: Condition (A) of the theorem holds for every interpretation if and only if it holds for every one-one interpretation.

A detailed definition and proof appear in [1].

Next, we have to prove some properties of how the $\cdot$ relation behaves for termination and initiation symbols.

Lemma 2: Let $\mathcal{S}$ be a persistent commutative lossless schema, $\mathcal{I}$ a one-one interpretation, and $\alpha = (c,q,\mu)$ an $\mathcal{I}$-instantaneous description. Then for each pair of operations a and b:

   (a) If $\alpha \cdot \bar{a}\, \bar{b}$ and $\alpha \cdot \bar{b}\, \bar{a}$ are defined then $\alpha \cdot \bar{a}\, \bar{b} = \alpha \cdot \bar{b}\, \bar{a}$

   (b) If $\alpha \cdot \bar{a}\, b_\ell$ and $\alpha \cdot b_\ell \bar{a}$ are defined then $\alpha \cdot \bar{a}\, b_\ell = \alpha \cdot b_\ell \bar{a}$ if and only if (i) $R(b) \cap D(a) = \emptyset$

      or (ii) b_ℓ is a repetition; i.e. $\Pi_{R(b)}(\alpha) = \Pi_{R(b)}(\alpha \cdot b_\ell)$.

(c) If $\alpha \cdot a_j b_\ell$ and $\alpha \cdot b_\ell a_j$ are defined then:

   (i) for $a \neq b$ $\alpha \cdot a_j b_\ell = \alpha \cdot b_\ell a_j$ if and only if $R(a) \cap R(b) = \emptyset$

   (ii) for $a = b$; $j = \ell$ and $\alpha \cdot a_\ell a_\ell = \alpha \cdot a_\ell a_\ell$.

Proof: The proof is by cases:

   (a) If $a = b$ then the result is obvious. $\mu(a)$ simply has two
       like objects added to it in each case.

If $a \neq b$ then $(c,q,\mu) \cdot \overline{a}\ \overline{b} = (c, \tau(q, \overline{a}\ \overline{b}), \mu')$

           and $(c,q,\mu) \cdot \overline{b}\ \overline{a} = (c, \tau(q, \overline{b}\ \overline{a}), \mu'')$,

and we wish to show that

   $(c, \tau(q, \overline{a}\ \overline{b}), \mu') = (c, \tau(q, \overline{b}\ \overline{a}), \mu'')$.

Since c is unchanged by initiations there is nothing to prove for
this component. Next $\tau(q, \overline{a}\ \overline{b}) = \tau(q, \overline{b}\ \overline{a})$ by commutativity.
Finally we compare μ' and μ''.

   For $d \neq a,b$ $\mu'(d) = \mu''(d) = \mu(d)$ $\mu'(a) = \mu(a)$, $\Pi_{D(a)}(c)$ and
$\mu''(a) = \mu(a)$, $\Pi_{D(a)}(c)$ so $\mu'(a) = \mu''(a)$ since initiations do not
modify memory. Similarly $\mu'(b) = \mu''(b)$ so part (a) of the lemma is
proved.

   (b) Let $a \neq b$ $(c,q,\mu) \cdot \overline{a}\ b_\ell = (c', \tau(q, \overline{a}\ b_\ell), \mu')$ and
       $(c,q,\mu) \cdot b_\ell \overline{a} = (c'', \tau(q, b_\ell \overline{a}), \mu'')$.

Here $c' = c''$ since the memory contents are not changed by $\overline{a}$, so in
each case termination b_ℓ is the only thing that changes the memory
contents. Next, by commutativity $\tau(q, \overline{a}\ b_\ell) = \tau(q, b_\ell \overline{a})$.

   Again we must check the μ lists.

   For $d \neq a,b$ $\mu'(d) = \mu''(d) = \mu(d)$. By definition $\overline{a}$ does not change
the μ list of b so $\mu'(b)$ is $\mu(b)$ with the first element deleted.
By definition $\mu''(b)$ is also $\mu(b)$ with the first element deleted. Thus
$\mu'(b) = \mu''(b)$. Now $\mu'(a) = \mu(a) \Pi_{D(a)}(c)$ and $\mu''(a) = \mu(a) \Pi_{D(a)}(c')$.

These are equal if and only if $\Pi_{D(a)}(c) = \Pi_{D(a)}(c')$.

Since the interpretation is one-one $\Pi_{D(a)}(c) = \Pi_{D(a)}(c')$ if and only if

$c = c'$ on $D(a)$. That is, $D(a) \cap R(b) = \emptyset$ or b_ℓ is a repetition.

Next, we must consider case (b) for $a = b$. For this case let

$(c,q,\mu) \cdot \bar{a} \, a_\ell = (c', \tau(q, \bar{a} \, a_\ell), \mu')$ and $(c,q,\mu) \cdot a_\ell \bar{a} = (c'', \tau(q, a_\ell \bar{a}), \mu'')$

$c' = c''$ since a_ℓ is the only termination and the $\bar{a}$ initiation does not

change the first element in the $\mu(a)$ list.

By commutativity $\tau(q, \bar{a} \, a_\ell) = \tau(q, a_\ell \bar{a})$. The μ lists μ' and μ''

can finally be shown equal by an argument essentially identical to that

for the case $a \neq b$.

(c) (ii) If $a = b$ then $\alpha \cdot a_j b_\ell$ and $\alpha \cdot b_\ell a_j$ are both defined only if

$\ell = j$, since the first outcome is uniquely determined by G_a. Thus

we have $\alpha \cdot a_j a_j$ and obviously this equals $\alpha \cdot a_j a_j$.

(c) (i) If $a \neq b$ let

$(c,q,\mu) \cdot a_j b_\ell = (c', \tau(q, a_j b_\ell), \mu')$

and $(c,q,\mu) \cdot b_\ell a_j = (c'', \tau(q, b_\ell a_j), \mu'')$

Using the one-one interpretation we see that $c' = c''$ if and only

if $R(b) \cap R(a) = \emptyset$. $\tau(q, q_j b_\ell) = \tau(q, b_\ell a_j)$ by commutativity, and

$\mu' = \mu''$ since in each case a single item is removed only from the $\mu(a)$

and $\mu(b)$ lists.

This completes the proof of the lemma.

Lemma 3: Let $\mathcal{S}$ be a persistent commutative lossless schema, $\mathcal{I}$ a one-one

interpretation, and α_0 the initial instantaneous description. Let

$v \in \Sigma^*$, and $\sigma, \pi \in \Sigma$ such that $\alpha_0 \cdot v\sigma\pi = \alpha_0 \cdot v\pi\sigma$. Then for any $w \in \Sigma^* \cup \Sigma^w$:

(a) $v\sigma\pi w$ is an $\mathcal{I}$-computation if and only if $v\pi\sigma w$ is.

(b) For any $i \in M$ $\Omega_i(v\sigma\pi w) = \Omega_i(v\pi\sigma w)$.

The proof of part (a) follows from the definition of the $\cdot$ relation, and part (b) follows by checking all the cases from Lemma 2.

Lemma 4: Let $\mathcal{S}$ be a persistent schema, $\mathcal{A}$ an interpretation and α_0 the initial instantaneous description. Let $u, v \epsilon \Sigma^*$, $w \epsilon \Sigma^* \cup \Sigma^\omega$ and $\sigma \epsilon \Sigma$.

(a) If $\alpha_0 \cdot u\sigma$ is defined, $\sigma \notin w$ and $\alpha_0 \cdot uv$ is defined, then

$\alpha_0 \cdot uv\sigma$ is defined.

(b) If $\alpha_0 \cdot u\sigma$ is defined and uw is an $\mathcal{A}$-computation then $\sigma \epsilon w$.

Part (a) follows from persistence, and part (b) from persistence and the finite delay property.

We now wish to prove the theorem. Suppose (A) holds, then we wish to prove that $\mathcal{S}$ is determinate. Assume $\mathcal{S}$ is not determinate. That is, there is an interpretation $\mathcal{A}$ such that x and y are $\mathcal{A}$-computations and for some $i \epsilon M$ $\Omega_i(x) \neq \Omega_i(y)$. We shall prove that for any $n \leq \ell(x)$ there is an $\mathcal{A}$-computation $z(n)$ such that:

(1) $z(n)$ has the same cell sequences as y,

i.e. $\Omega_i z(n) = \Omega_i(y)$ for all $i \epsilon M$.

(2) $_n(z(n)) = {}_n x$.

Since this is true for all $n \leq \ell(x)$ we obtain $_n(\Omega_i(z(n))) = {}_n(\Omega_i(x))$ $= {}_n(\Omega_i(y))$ so for no n can $\Omega_i(x)$ and $\Omega(y)$ differ. This provides a contradiction showing that $\Omega_i(x) = \Omega_i(y)$.

The proof of these two properties of $z(n)$ is done inductively on n.

Basis: Assume $z(0) = y$ so (1) and (2) hold for $n = 0$.

Inductive Assumption: Assume (1) and (2) hold for $n = k$ and $\ell(x) \geq k+1$ (otherwise we are done). Then $\alpha_0 \cdot (_k x) x_{k+1}$ is defined so $\alpha_0 \cdot t \; x_{k+1}$ is defined, where $t = {}_k(z(k))$, and $\alpha_0 \cdot (_k x) x_{k+1} = \alpha_0 \cdot t \; x_{k+1}$ since by the inductive assumption (2) holds for $n = k$. By Lemma 4 $z(k) = tvx_{k+1} u$, $x_{k+1} \notin v$. That is, x_{k+1} appears somewhere in the sequence since $\alpha_0 \cdot tx_{k+1}$ is defined. If v is null then let

$z(k+1) = z(k)$ and (1) and (2) hold. If v is not null then

$v = w\pi$, $w\epsilon\Sigma^*$ $\pi\epsilon\Sigma$. Since $\alpha_0 \cdot tx_{k+1}$ is defined it follows from (a) of

Lemma 4 that $\alpha_0 \cdot twx_{k+1}$ and $\alpha_0 \cdot tw\pi x_{k+1}$ are defined. Also, since

$\alpha_0 \cdot tw\pi$ is defined, $\alpha_0 \cdot twx_{k+1}\pi$ is defined. But by the assumption that

(A) holds

$$\alpha_0 \cdot tw\pi x_{k+1} = \alpha_0 \cdot twx_{k+1}\pi$$

Thus, by Lemma 3, $twx_{k+1}\pi u$ is an $\mathscr{A}$-computation, and it has the same cell

contents sequences as $z(k)$, that is, for all i, $\Omega_i(z(k)) = \Omega_i(twx_{k+1}\pi u)$.

Now, by identical reasoning we can continue to "slide" x_{k+1} to the

left until we have $tx_{k+1}vu$ with the same cell contents sequences as

$z(k)$ (and therefore the same as y). We call this $z(k+1)$ and note that

it satisfies (1) and (2). This accomplishes the goal of proving that

(A) $\Longrightarrow$ determinacy. (The "sliding argument" is a technique which is

central to a number of proofs about schema properties.)

The remainder of the proof of the theorem is to show that determinacy

implies condition (A). Assume determinacy but (A) does not hold, that is,

$\alpha_0 \cdot u\sigma\pi \neq \alpha_0 \cdot u\pi\sigma$. Then by cases of Lemma 2 one can show a difference in

$\Omega_i(u\sigma\pi)$ and $\Omega_i(u\pi\sigma)$ for some i, but this contradicts determinacy,

completing the proof of the theorem.

This theorem shows how determinacy, which is a property on sequences

of values in memory cells, is equivalent to a type of "commutativity" of

events in forming instantaneous descriptions. That is, when a "racing"

of several operation performances does not cause a change in behavior.

Decidability of Determinacy

We now wish to show that, for a special class of schemata, determinacy

is decidable. To do this, we must first digress to a combinatorial structure called vector addition systems, show how a certain property of these systems is decidable, and then show the connection of this to schemata.

<u>Definition</u>: An r-dimensional <u>vector addition system</u> is a pair $\mathcal{W} = (d,W)$
where d is an r-dimensional vector of nonnegative integers, and W is a finite set of r-dimensional integer vectors.

The <u>reachability set</u> $R(\mathcal{W})$ is the set of points in the first orthant that can be reached from d by successively adding vectors in W such that the path of points so formed always remains in the first orthant. As a simple example considers $r = 2$, $d = (1,1)$ and $W = \{(-2,1), (0,1), (3,-1)\}$. Note, for example that $(4,2)\epsilon R(\ )$.

$$(4,2) = (1,1) + (3,-1) + (0,1) + (0,1)$$

and the successive points $(4,0)$, $(4,1)$, $(4,2)$ are all in the first orthant.

Some elementary terminology is needed. For r-vectors x and y $x \leq y$ if and only if $x_i \leq y_i$, $i = 1,2,\cdots,r$. We use 0 to denote the vector of all zeros when appropriate, and ω is a symbol for which, for any integer n, $n + \omega = \omega$. A <u>rooted tree</u> is a tree with some designated vertex called the <u>root</u>. All vertices but the root have one edge directed into the vertex. If ξ and η are vertices of a rooted tree with a directed path from ξ to η, then we designate this by $\xi \prec \eta$. If there is a directed edge from ξ to η then η is called a <u>successor</u> of ξ. If η is a node with no edge directed out then η is called an <u>end</u>.

For $\mathcal{W}$ we construct a tree $\mathcal{J}(\mathcal{W})$ with labelled vertices $\ell(\xi)$ for ξ, where $\ell(\xi)$ is an r-dimensional vector label having components from $N \cup \{\omega\}$.

<u>Definition</u>: $\mathcal{J}(\mathcal{W})$ consists of:
(1) root δ with $\ell(\delta) = d$.

(2) Let η be a vertex of $\mathcal{T}(\mathcal{W})$

(a) if, for some vertex $\xi \prec \eta$ $\ell(\xi) = \ell(\eta)$ then η is an end.

(b) otherwise successors of η are formed (one for each $w \in W$

for which $\ell(\eta) + \dot{w} \geq 0$.

Let η_w denote the successor of η associated with $w \in W$, then $\ell(\eta_w)$

is determined as follows:

(i) if there is a $\xi \prec \eta_w$ such that $\ell(\xi) \leq \ell(\eta) + w$ and

$(\ell(\xi))_i < (\ell(\eta) + w)_i$ then $(\ell(\eta_w))_i = \omega$.

(ii) if no such ξ exists, then $(\ell(\eta_w))_i = (\ell(\eta) + w)_i$.

We demonstrate the $\mathcal{T}(\mathcal{W})$ construction so defined using our previous

example $\mathcal{W}$. $\mathcal{T}(\mathcal{W})$ is:

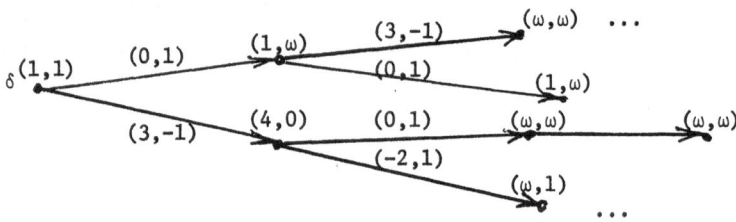

Here only a starting portion of the tree is given and edges are labelled

with the elements in W used to form the successor vertex.

<u>Theorem</u>:   For any vector addition system $\mathcal{W}$, $\mathcal{T}(\mathcal{W})$ is finite.

<u>Proof</u>:   Assume $\mathcal{T}(\mathcal{W})$ contains an infinite path from its root

$\delta, \eta_1, \eta_2, \cdots$.   Then $\ell(\delta), \ell(\eta_1), \ell(\eta_2), \cdots$ is an infinite

sequence of labels with elements from $N \cup \{\omega\}$. This must contain

an infinite subsequence of labels $\ell(\eta_{i_1}), \ell(\eta_{i_2}), \cdots; i_1 < i_2 \cdots$ such

that $\ell(\eta_{i_1}) \leq \ell(\eta_{i_2}) \leq \cdots$.   Now, none of these vertices is an

end, therefore, we never have $\ell(\eta_{i_j}) = \ell(\eta_{i_{j+1}})$.   Thus $\ell(\eta_{i_{j+1}})$

has at least one coordinate greater than $\ell(\eta_{i_j})$ and by the definition of $\mathcal{J}(\mathcal{U})$ this coordinate must equal ω. Now, since the number of coordinates is finite we see that no such infinite path can exist. Also, we know that the number of edges leaving any vertex of $\mathcal{J}(\mathcal{U})$ is finite -- actually $\leq |W|$. König's lemma now says that $\mathcal{J}(\mathcal{U})$ is finite, completing the proof.

The form of König's lemma we use is:

Let T be a rooted tree in which each vertex has only a finite number of successors and there is no infinite path directed away from the root. Then T is finite.

This theorem, with the recursive definition of $\mathcal{J}(\mathcal{U})$, gives us an algorithm for constructing $\mathcal{J}(\mathcal{U})$.

Theorem: Let x be a nonnegative integer vector, then the following are

    equivalent:

    (1) there is a $y \epsilon R(\mathcal{U})$ such that $x \leq y$.

    (2) there is a vertex $\eta \epsilon \mathcal{J}(\mathcal{U})$ such that $x \leq \ell(\eta)$.

This is a central decidability result of vector addition systems which is useful for schemata property decidability. A proof can be found in [1]. A corollary of this theorem states that it is decidable for any vector addition system $\mathcal{U}$, and point $x \geq 0$, whether there exists a $y \epsilon R(\mathcal{U})$ such that $y \geq x$.

We are now ready to look at the connection between vector addition systems and schemata.

Definition: A counter schema $\mathcal{S} = (M, A, \mathcal{J})$ has $\mathcal{J}$ defined by:

    (1) a nonnegative integer k, the number of "counters"

    (2) a finite set Σ

    (3) a finite set S with a distinguished element s_0.

    (4) a vector $\pi \epsilon N^k$

R. E. Miller

(5)   a function v from Σ into N^k such that if $\sigma \epsilon \Sigma_t$ then

$v(\sigma) \geq 0$.

(6)   a partial function $\theta : S \times \Sigma \longrightarrow S$ which is total on $S \times \Sigma_t$.

Then: $\mathcal{J} = (Q, q_0, \Sigma, \tau)$ where:

$Q = S \times N^k$, $q_0 = (s_0, \pi)$, $\tau((s,x),\sigma)$ is defined if $\theta(s,\sigma)$ is defined and $x + v(\sigma) \geq 0$, and in this case $\tau((s,x),\sigma) = (\theta(s,\sigma), x + v(\sigma))$.

Thus, the control for a counter schema consists of a finite set of states plus a finite set of counters, and it is these state-counter pairs that act as states in Q for the control. Each initiation or termination causes a change in the S part of the state and a change in the counter values.

We can connect counter schemata and vector addition systems by constructing a vector addition system for any counter schema such that the vector addition system in some sense simulates the control of the schema.

For a counter schema $\mathcal{J}$ we construct a vector addition system $\mathcal{W}_\mathcal{J}$ with coordinates

$$|S| + k + |A|$$

where the $|S|$ coordinates represent the S state behavior, the k coordinates directly represent counter values, the $|A|$ coordinates represent μ list lengths for the operations in A.

We define the d part of the $\mathcal{W}_\mathcal{J} = (d,W)$ as follows:

$d(s_0) = 1$

$d(s) = 0 \quad s \epsilon S, s \neq s_0$

$d(i) = \pi_i \quad i = 1, 2, \cdots, k$

$d(a) = 0 \quad a \epsilon A$

The W part we describe by looking separately at the form of vectors

R. E. Miller

for each part. The $|S|$ part has vectors of the form:

$$0 \cdots 0 \ 1 \ 0 \cdots 0 \ -1 \ 0 \cdots 0$$

having one $+1$ (say in the ith position) and one -1 (say in the jth position) for each case in which there is some σ such that $\theta(s_j, \sigma) = s_i$
For the k part, for each σ we put the $v(\sigma)$ values in the proper coordinates.

For the $|A|$ part, if σ is on initiation put an $a+1$ in the coordinate associated with the operation designated by σ, otherwise the value is unchanged. If σ is a termination of some operation a, then the A part vector has a -1 in the a coordinate.

With this description we should note that

(1) The state the schema is in is denoted by a one in some coordinate of the $|S|$ part of the partial sum vector. All other $|S|$ part coordinates are zero.

(2) The k coordinates indicate the current counter values.

(3) In the $|A|$ part initiations can only occur if defined and terminations can only occur if the μ list value -- indicated by the coordinate value -- is >0.

Thus, proceeding from reachable point to reachable point in $R(\mathcal{U}_{\mathcal{S}})$ corresponds to a computation for $\mathcal{S}$.

We now can state the decidability theorem of interest to us.

Theorem: It is decidable whether a repetition-free lossless persistent commutative counter schema is determinate.

The proof of this theorem is based on the constructibility of $\mathcal{J}(\mathcal{U}_{\mathcal{S}})$ and the necessary and sufficient properties for determinacy stated in a previous theorem and its proof. Essentially we argue as follows: The lack of determinacy was shown to arise from cases in which one had a race

between two operations a and b for which $D(a) \cap R(b) \neq \emptyset$, and the
performances in question were not repetitions or where $R(a) \cap R(b) \neq \emptyset$.
Here, since we assume repetition-freeness, we need not concern ourselves
with repetitions. As an aside, in [1] it is shown that repetition-freeness
is itself a decidable property for schemata. Thus, we need only check if
we have any "races" between operation pairs (a,b) where $D(a) \cap R(b) \neq \emptyset$
or $R(a) \cap R(b) \neq \emptyset$. But, since A is finite, there are only a finite
number of such pairs to check. Finally, since $\mathcal{J}(\mathcal{U}_{\mathcal{S}})$ can be constructed,
and this will indicate pairs of operations that can be concurrently in
operation by having μ list values simultaneously greater than zero, each
such pair can be tested so determinacy can be decided.

Undecidability of Determinacy and Other Properties

In the previous lecture we have shown that it is decidable whether a
repetition-free lossless persistent commutative counter schema is determinate. In this lecture we show that determinacy becomes undecidable when the
repetition-free part of the hypothesis is deleted. We show this by a
schema construction which shows that the decidability of determinacy under
these conditions would imply the decidability of the Post correspondence
problem, a well-known undecidable problem.

The form of Post correspondence problem we use is as follows:
Given two n-tuples of words

$$X = x_1, x_2, \cdots, x_n$$
$$Y = y_1, y_2, \cdots, y_n$$

over the alphabet $\{b_1, b_2\}$ it is undecidable whether there exists a
sequence of indices $i_1, \cdots, i_p$ such that:

$$x_{i_1} \cdots x_{i_p} = y_{i_1} \cdots y_{i_p}.$$

We begin our schema construction with a construction of $\mathcal{S}(X)$ which starts in state q_0 and ends in state q_e if and only if the computation has the form

$$\underbrace{a_1^{i_1-1} a_2 x_{i_1}}_{i_1 a's} \underbrace{a_1^{i_2-1} a_2 x_{i_2}}_{i_2 a's} \cdots x_{i_p} a_3 b_3$$

In $\mathcal{S}(X)$ $A = \{a,b\}$ $D(a) = R(a) = \{1\}$

$$D(b) = R(b) = \{2\}$$

$$K(a) = K(b) = 3$$

Since a and b operate on different memory locations they do not interact with each other. Thus the sequence of outcomes is determined only by the interpretation.

An example schema $\mathcal{S}(X)$ for $X = x_1, x_2$ where $x_1 = b_2 b_1$ and $x_2 = b_2 b_2 b_1$ is shown below.

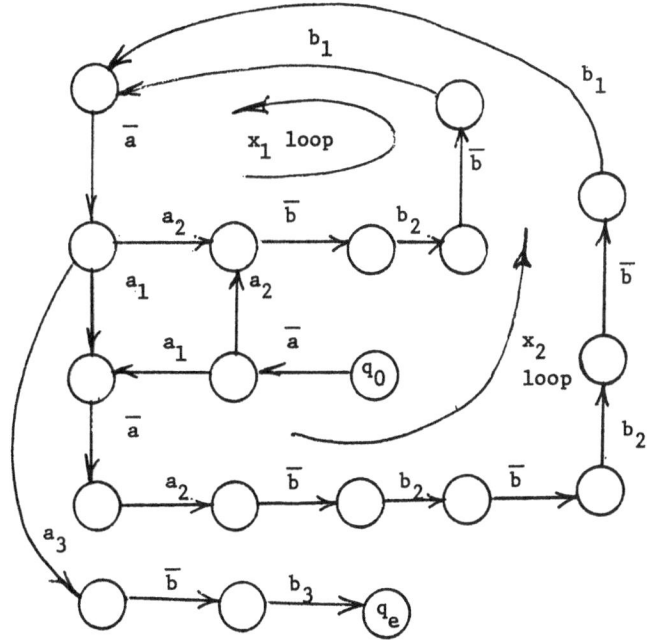

Note that the sequence of outcomes of a's determine whether x_1 or x_2 will occur.

R. E. Miller

Using $\mathcal{S}(X)$ and a similarly constructed $\mathcal{S}(Y)$ for Y we construct a schema $\mathcal{S}(XY)$, by adding an operation r, as follows:

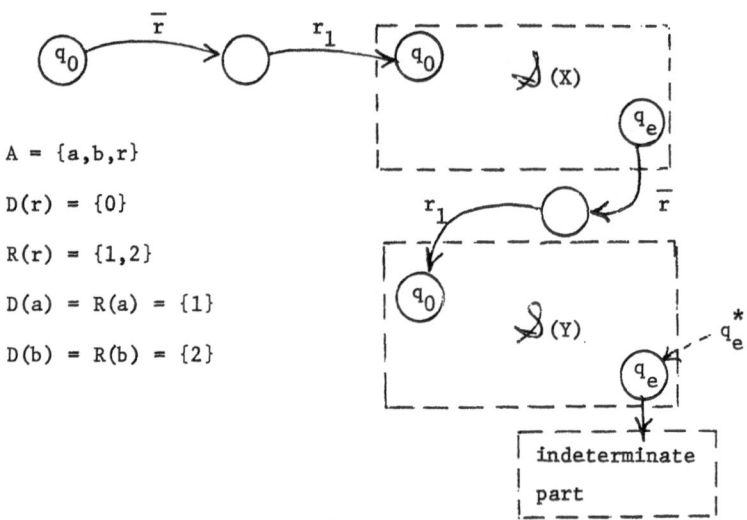

A = {a,b,r}

D(r) = {0}

R(r) = {1,2}

D(a) = R(a) = {1}

D(b) = R(b) = {2}

It is readily seen that state q_e^* is reached if and only if there is a solution to the Post correspondence problem. Operation r sets locations 1 and 2 to certain values when entering q_0 of $\mathcal{S}(X)$ then if q_e of $\mathcal{S}(X)$ is reached operation r resets locations 1 and 2 to the same values as when entering $\mathcal{S}(X)$. Thus the same sequence of a outcomes will occur and this will reach q_e^* if and only if we get matching sequences from X and Y.

A simple indeterminate part, to add on after q_e^* is:

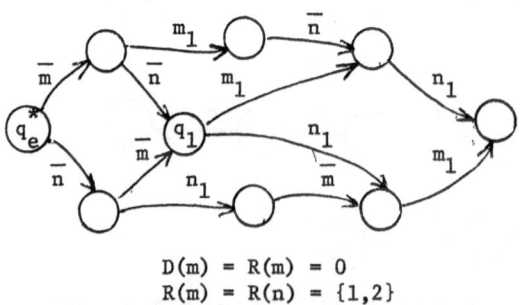

$$D(m) = R(m) = 0$$
$$R(m) = R(n) = \{1,2\}$$

R. E. Miller

The sequences of values in {1,2} thus depend upon the order of termination of m and n, and this provides the desired indeterminacy. Thus, clearly, this schema is indeterminate if and only if there is a solution to the Post correspondence problem, so determinacy is undecidable. We thus have our desired result.

Theorem: It is undecidable whether a lossless persistent commutative
counter schema is determinate.

The schema $\overset{3}{\lessgtr}$ (XY) satisfies all the hypotheses of the theorem, in fact it is somewhat more restrictive in form. Thus we see that the property of repetition-freeness provides a sharp boundary between decidability and undecidability of determinacy.

Other properties of schemata can be easily shown to be undecidable by adding on a part after q_e^* of $\overset{3}{\lessgtr}$ (XY) that violates (or tests) the property of interest. A simple example is to determine if some operation of a schema is ever performed. This can be seen undecidable simply by adding operation transitions for a new operation after q_e^*. Other properties so testable are: Boundedness, serial operation, termination, and reachability of a state.

Composition and Renamings

In this lecture we briefly describe the work of references [3,4,5]. The idea of schema composition goes back to the original notion that a parallel program schema is a model for a parallel program. If one wishes to develop a large program, then it is often convenient to first develop identifiable subtasks as subprograms and then later put these subprograms together in a suitable fashion to make the complete program. Similarly, to model the program it might be desirable to first model certain subprograms by schemata then later to "compose" these schemata together

suitably to represent the complete program. This could be viewed as a
simple structured approach to modelling the program. The work on schema
composition is aimed at defining some basic types of interconnections for
schemata and then proving theorems that say, essentially, that the proper
behavior of the subparts is carried over to the complete schema when
composition is done in the prescribed way.

In [3,4] special types of schemata, called finishing schemata and exit
schemata, are defined which are suitable for defining composition. We omit
the details of these schema definitions here, but just point out some of
their essential features. In both cases these schemata are assumed to have
a finite subset of "begin" states and a finite subset of "end" states. The
begin and end states are useful in composition, as we shall see. For end
states, one assumes that no transitions are defined out of end states, and
that whenever an end state is reached a finite computation for the schema
has been completed. Additional constraints upon finishing schemata result
in the fact that no more than one performance of any operation can be going
on simultaneously. This restricts μ lists to be either of length zero
or one and simplifies the analysis. From this point on, for composition,
when we say schemata we will mean finishing schemata.

We will now define four types of composition and state some results
about the composed forms. We use terminology that subscripts symbols with
the schema symbol to avoid confusion. Thus, for example, if we have two
schemata $\mathcal{A}$ and $\mathcal{B}$ we let $\mathcal{A} = (M_\mathcal{A}, A_\mathcal{A}, \mathcal{J}_\mathcal{A})$ and $\mathcal{B} = (M_\mathcal{B}, A_\mathcal{B}, \mathcal{J}_\mathcal{B})$,
etc.

The <u>serial</u> <u>composition</u> of two schemata $\mathcal{A}$ and $\mathcal{B}$ is designated by
$0(\mathcal{A}, \mathcal{B})$. For this composition we require $Q_\mathcal{A} \cap Q_\mathcal{B} = \emptyset$ and that the
number of end states of $\mathcal{A}$ equal the number of begin states of $\mathcal{B}$. Essen-

R. E. Miller

tially $0(\mathcal{A}, \mathcal{B})$ is a serial linking from end states of $\mathcal{A}$ to destinations of begin states of $\mathcal{B}$. Let $\mathcal{A} = (M_{\mathcal{A}}, A_{\mathcal{A}}, \mathcal{T}_{\mathcal{A}})$ and $\mathcal{B} = (M_{\mathcal{B}}, A_{\mathcal{B}}, \mathcal{T}_{\mathcal{B}})$ with the end states of $\mathcal{A}$ being $E_{\mathcal{A}} = \{e_1, e_2, \cdots, e_n\}$ and the begin states of $\mathcal{B}$ being $B_{\mathcal{B}} = \{b_1, b_2, \cdots, b_n\}$. Then we define $0(\mathcal{A}, \mathcal{B}) = (M, A, \mathcal{T})$ where:

$$M = M_{\mathcal{A}} \cup M_{\mathcal{B}}$$
$$A = A_{\mathcal{A}} \cup A_{\mathcal{B}},$$
$$\Sigma = \Sigma_{\mathcal{A}} \cup \Sigma_{\mathcal{B}}$$
$$Q = Q_{\mathcal{A}} \cup Q_{\mathcal{B}}$$
$$E = E_{\mathcal{B}}$$
$$B = B_{\mathcal{A}}.$$

The transition function τ is defined by:

$$\tau(q,\sigma) = \begin{cases} \tau_{\mathcal{A}}(q,\sigma) & \text{if } \tau_{\mathcal{A}}(q,\sigma) \text{ is defined.} \\ \tau_{\mathcal{B}}(q,\sigma) & \text{if } \tau_{\mathcal{B}}(q,\sigma) \text{ is defined} \\ \tau_{\mathcal{B}}(b_i,\sigma) & \text{if } \tau_{\mathcal{B}}(b_i,\sigma) \text{ is defined and } q = e_i. \end{cases}$$

Pictorially the serial composition can be shown as follows:

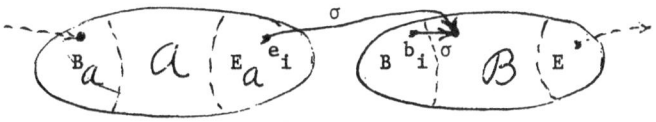

Intuitively this composition makes end state e_1 of $\mathcal{A}$ act like begin state b_1 of $\mathcal{B}$ for a computation of $\mathcal{A}$ ending in e_1, and then a computation of $\mathcal{B}$, starting in b_1 could follow.

The <u>concurrent</u> <u>composition</u> of $\mathcal{A}$ and $\mathcal{B}$, denoted by $X(\mathcal{A}, \mathcal{B})$ is defined when $Q_{\mathcal{A}} \cap Q_{\mathcal{B}} = \emptyset$ and $A_{\mathcal{A}} \cap A_{\mathcal{B}} = \emptyset$, as $X(\mathcal{A}, \mathcal{B}) = (M, A, \mathcal{T})$ where:

$$M = M_{\mathcal{A}} \cup M_{\mathcal{B}}$$
$$A = A_{\mathcal{A}} \cup A_{\mathcal{B}}$$
$$\Sigma = \Sigma_{\mathcal{A}} \cup \Sigma_{\mathcal{B}}$$

R. E. Miller

states are pairs, $Q = Q_{\mathcal{A}} \times Q_{\mathcal{B}}$

$$B = B_{\mathcal{A}} \times B_{\mathcal{B}}$$

$$E = E_{\mathcal{A}} \times E_{\mathcal{B}}$$

The transition function is defined only in the following cases (let

$q_{ij} = (q_i)_{\mathcal{A}} \times (q_j)_{\mathcal{B}}$)

If $\sigma \epsilon \Sigma_{\mathcal{A}}$ then $\tau(q_{ij}, \sigma) = (\tau_{\mathcal{A}}(q_i, \sigma), q_j)$

If $\sigma \epsilon \Sigma_{\mathcal{B}}$ then $\tau(q_{ij}, \sigma) = (q_i, \tau_{\mathcal{B}}(q_j, \sigma))$

The concurrent composition of two schemata allows parallel operation

of the two schemata in a simple FORK-JOIN manner. Pictorally it can be

viewed as:

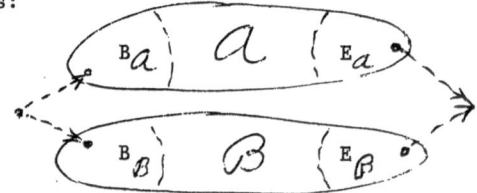

The first type of result needed, is that under these compositions the

type of object we get is still within the class of objects we wish to

study. This is so.

<u>Theorem</u>: If $\mathcal{A}$ and $\mathcal{B}$ are schemata, then $0(\mathcal{A}, \mathcal{B})$ and $X(\mathcal{A}, \mathcal{B})$ are

schemata.

This is a basic closure result. The next types of results deal with

the form of $\mathcal{A}$-computations for the composed schemata. Without going

into details, interpretations for the composed form are formed from

"compatible" interpretations from the constituent schemata. Compatible

here means that if the two interpretations are defining something for the

same element, then the definitions are the same.

<u>Theorem</u>: Let $\mathcal{A}$ and $\mathcal{B}$ be schemata with $0(\mathcal{A}, \mathcal{B})$ defined. Any compu-

tation z of $0(\mathcal{A}, \mathcal{B})$ is of one of the following two forms:

R. E. Miller

(1) z = x, where x is a nonterminating computation of $\mathcal{A}$.

(2) z = xy where x is a terminating computation of $\mathcal{A}$ and y

is a computation of $\mathcal{B}$.

To state a similar result for concurrent composition we first need to define a "memory conflict" relation between $\mathcal{A}$ and $\mathcal{B}$.

Let $D_{\mathcal{A}} = \bigcup_{a \in A_{\mathcal{A}}} D(a)$ and $R_{\mathcal{A}} = \bigcup_{a \in A_{\mathcal{A}}} R(a)$, and let $D_{\mathcal{B}}$ and $R_{\mathcal{B}}$ be similarly defined. Then we say

$$\mathcal{A} \, \rho \, \mathcal{B} \Longleftrightarrow [D_{\mathcal{A}} \cap R_{\mathcal{B}}] \cup [D_{\mathcal{B}} \cap R_{\mathcal{A}}] \cup [R_{\mathcal{A}} \cap R_{\mathcal{B}}] \neq \emptyset.$$

Theorem: If $\mathcal{A}$ and $\mathcal{B}$ are schemata such that $X(\mathcal{A},\mathcal{B})$ is defined and

$\mathcal{A} \, \not{\rho} \, \mathcal{B}$, then any computation of $X(\mathcal{A},\mathcal{B})$ is a shuffle of a

computation of $\mathcal{A}$ and a computation of $\mathcal{B}$.

The "shuffle" is essentially combining two strings into a single string, with the order of each original string maintained in the combined string, but having no other restrictions on the number of contiguous symbols (other than it being finite) taken from one string, before one or more symbols are taken from the other string.

Another form of composition consists of connecting an end state e of a schema to a begin state b to form a loop. We define such a composition, which we call an __iterate__ and designate this for a schema $\mathcal{A} = (M_{\mathcal{A}}, A_{\mathcal{A}}, \mathcal{T}_{\mathcal{A}})$ as $+(\mathcal{A}, e, b) = (M, A, \mathcal{T})$ where $M = M_{\mathcal{A}}$, $A = A_{\mathcal{A}}$, $\Sigma = \Sigma_{\mathcal{A}}$, $Q = Q_{\mathcal{A}}$, and $B = B_{\mathcal{A}}$. $E = E_{\mathcal{A}} - \{e\}$ and τ for $+(\mathcal{A}, e, b)$ is defined as follows:

$$\tau(q,\sigma) = \begin{cases} \tau_{\mathcal{A}}(b,\sigma) & \text{if } q = e \text{ and } \tau_{\mathcal{A}}(b,\sigma) \text{ is defined} \\ \tau_{\mathcal{A}}(q,\sigma) & \text{whenever } \tau_{\mathcal{A}}(q,\sigma) \text{ is defined.} \end{cases}$$

Clearly $+(\mathcal{A}, e, b)$ is a schema, so we also get the desired closure result for this type of composition. The computations of $+(\mathcal{A}, e, b)$ are characterized by the next theorem.

Theorem: Let $\mathcal{a}$ be a schema. If z is a computation for $+(\mathcal{a}, e, b)$ then

(1) $z = x^1 x^2 \cdots,$ where x^i are computations for $\mathcal{a}$ ending in state e, or

(2) $z = x^1 x^2 \cdots x^k y$ where $k \geq 0$ and for all $i = 1, 2, \cdots k$ x^i are computations for $\mathcal{a}$ ending in e and y is a computation for $\mathcal{a}$ not ending in e.

The final type of composition we consider is much more complex. We give only an intuitive idea of the form of this composition ([4] contains details). The idea here is that we would like to replace an operation as defined in one schema by a more detailed description (i.e. a schema) of the operation. This type of composition we call insertion. It allows one to hierarchically define a program. Naturally, a number of consistency requirements must be satisfied, such as the number of outcomes of the operation matching, in some sense, the number of end states of the schema replacing the operation. In [4] closure and computation characterization theorems are given for insertion.

Another class of theorems is given in [3,4] for composition. These theorems show under what conditions determinacy carries over from the constituent schemata to the composed schemata. Briefly, this usually involves some constraints upon the operations and their effect on memory locations.

For our schemata which we have defined and studied so far we have always specified the memory locations $D(a)$ and $R(a)$ that an operation a effects when it is performed. However, it may be advantageous to respecify the $D(a)$ or $R(a)$ locations of some operations. For example, some operation might be storing a "temporary result" in some location which is later used as an operand for another operation. Because this location is used during this period, however, it may restrict the use of this location by other operations, and only due to this memory conflict it may mean that these other operations must wait until the first pair of operations have finished using the location in question. This situation illustrates that by a reallocation of memory, or a "renaming," we could attain more parallelism. Secondly, a renaming might decrease the number of memory locations needed to perform a computation, thus providing an economy of memory usage. This then is the subject of "renaming" in schemata. How can the assignment of memory locations be changed in a consistent and advantageous way? We will illustrate this notion of renaming (as done in [5]) only by two simple examples. The first example considers only a simple sequence of function evaluations -- or a computation. The sequence is:

$$(f(2,3) \rightarrow 0), \ (g(3) \rightarrow 1), \ (h(0,1) \rightarrow 0,3), \ (m(3) \rightarrow 3)$$

That is, we start by performing a function evaluation f which uses locations 2 and 3 for operands and places a result in location 0, then follow this by a g using 3 and putting its result in location 1, etc.

Here we note that the result of the f calculation (in location 0) is used by h. That is, location 0 is busy for this usage over the segment indicated below:

$$(f(2,3) \rightarrow 0), \ (g(3) \rightarrow 1), \ (h(0,1) \rightarrow 0,3), \ (m(3) \rightarrow 3)$$

  0-busy   3-busy

Similarly 3 is in use for a certain purpose where shown.

Now this use of location 0 could be moved to a different location. For example, no change in the final results in locations 0,1,2 or 3 would result if we changed this use to location 4 rather than 0, i.e. giving

$$(f(2,3) \rightarrow 4), \ (g(3) \rightarrow 1), \ (h(4,1) \rightarrow 0,3), \ (m(3) \rightarrow 3).$$

R. E. Miller

This renaming works but seems to be of no great use. However, if location 2 were of no interest beyond the use in the first f calculation we could rename the use of 0 to a use of 2 giving:

(f(2,3) → 2), (g(3) → 1), (h(2,1) → 0,3), (m(3) → 3)

and this would "free" location 0 for a longer period of time. The reader may see that the use of location 3 indicated above could also be changed by a renaming. Thus, for computation sequences, we are interested in contiguous segments of location usage starting with the point a value is stored in the location, and ending with the last usage of that value. This is one of the ideas developed by Logrippo in [5]. This idea of renamings can now be extended to schemata. We use here a different representation of schemata which looks more like a flowchart.

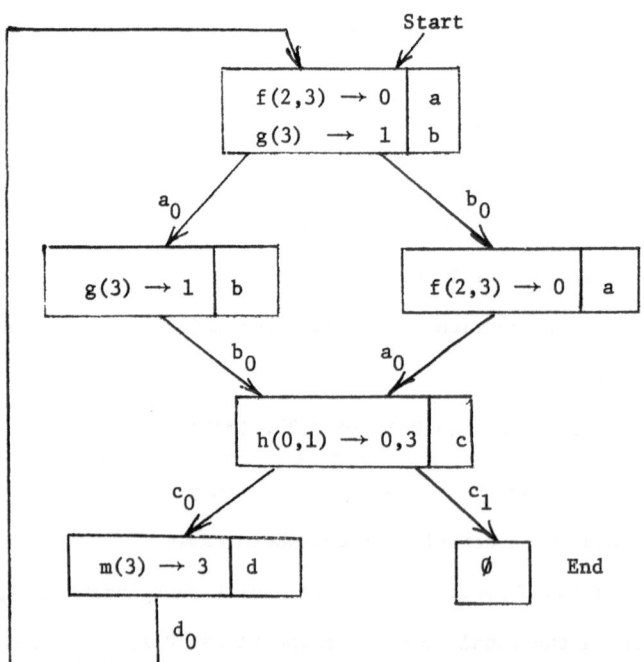

In this case each box of the flowchart specifies function operations on memory a refers to operation a which is f(2,3) → 0, b refers to

$g(3) \rightarrow 1$, c to $h(0,1) \rightarrow 0,3$, and d to $m(3) \rightarrow 3$. More than one operation in a box indicates concurrent performance is possible. Arrows are labelled with operation outcomes controlling the flow through the schema flowchart. Note here, that the computation we looked at earlier is the start of one possible computation in this schema.

Now, we are again interested in renamings which do not change the overall behavior. Rather than simple segments, we now get "regions" of usage for each usage of each location. The regions for location 3 are shown in the next figure.

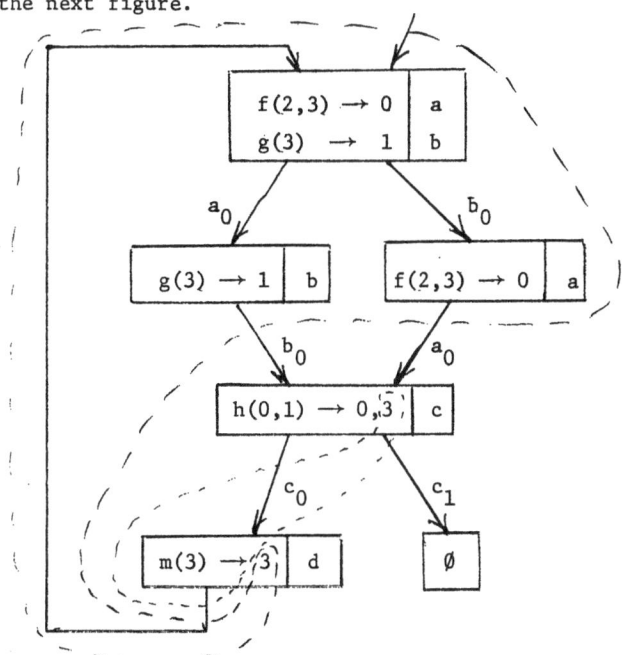

Clearly, for consistent renamings the renaming must be done consistently through a complete region so outlined.

The properties of such regions for renaming, and how they can be used advantageously for maximizing parallelism or for conserving memory are studied in [5] and related papers.

REFERENCES

[1] R. M. Karp and R. E. Miller, "Parallel Program Schemata," Journal of
     Computer and Systems Sciences, Vol. 3, May 1969, pp. 147-195.

[2] R. E. Miller, "Some Undecidability Results for Parallel Program
     Schemata," SIAM Journal on Computing, Vol. 1, No. 1, March 1972,
     pp. 119-129.

[3] W. A. Brinsfield and R. E. Miller, "On the Composition of Parallel
     Program Schemata," Conference Record, 1971, 12th Annual Symposium
     on Switching and Automata Theory, October 1971, pp. 20-23.

[4] W. A. Brinsfield and R. E. Miller, "Insertion of Parallel Program
     Schemata," Proceedings of the Seventh Annual Princeton Conference
     on Information Sciences and Systems, March 1973.

[5] L. Logrippo, "Renamings in Parallel Program Schemas," Research Report
     TR-74-01, November 1974, University of Ottawa, Department of
     Computer Science, Ottawa, Canada.

R. E. Miller

Relationships between Various Models of Parallelism and Synchronization

R. E. Miller

Introduction

There has been considerable work on many different facets of parallel computation in recent years (surveys on some of these are given in [1], [8] and [9]). Since the early 1960's there have been many different formal models studied as representations for parallel computation, cooperating processes, deadlock situations and multiprocessing. We wish to show here how some of these models are actually very closely related to each other.

Also, we illustrate how developing these relationships enables one to carry over results about the properties of one model to results about properties of other models. A great deal of care is needed in doing these comparisons. First, a precise notion of "represented by" is needed in each case. Second, one must make sure that the properties of interest in one model actually translate to properties of interest in the other model; and lastly, one must make sure that the resulting "behavior" also carries over in a meaningful way. We illustrate these problems in the relationships we develop.

The models we will discuss are computation graphs [6], vector addition and vector replacement systems [5,6], Petri nets and generalized Petri nets [2,4,5,8,10], and semaphore systems [3]. Details of these comparisons can be found in [8] and [9].

Computation graphs and Petri nets were defined in an earlier lecture in which it was shown how marked graphs could be represented by computation graphs. We use, but do not repeat, these definitions again in this lecture.

R. E. Miller

Vector Addition and Vector Replacement Systems

Vector addition systems were introduced in [7] to study properties of schemata. The generalization to vector replacement systems appears in [5].

A <u>vector</u> <u>replacement</u> system $\mathcal{V} = (d,V)$ consists of (1) d, an r-dimensional vector of non-negative integers and, (2) V, a finite set of ordered pairs of r-dimensional integer vectors $V = \{(u_1,v_1),(u_2,v_2), \ldots,(u_p,v_p)\}$ where $u_i \leq 0$ and $u_i \leq v_i$, $i = 1,2,\ldots,p$.

The <u>reachability set</u> $R(\mathcal{V})$ is the set of all vectors of the form:

$$d + v^{(1)} + v^{(2)} + \ldots + v^{(s)}$$

such that $v^{(j)}$ is some v_i for each j, and

$$d + v^{(1)} + v^{(2)} + \ldots + v^{(k-1)} + u^{(k)} \geq 0$$

for all $k = 1,2,\ldots,s$, where we use the terminology: If $v^{(k)} = v_i$ then $u^{(k)} = u_i$.

A <u>vector</u> <u>addition</u> <u>system</u> is simply a vector replacement system for which whenever a jth component $(u_i)_j < 0$ then $(u_i)_j = (v_i)_j$. For reachability questions it readily follows that in vector addition systems the u_i vectors are dispensable.

Vector Replacement Systems and Computation Graphs

Let G be a computation graph with ℓ nodes and t edges, and let $I(n_i)$ denote the set of edge indices directed into node n_i, and $O(n_i)$ the set of edge indices directed out of node n_i.

Define $\mathcal{V}(G) = (d,V)$ as:

$$d = (A_1,A_2,\ldots,A_t) \ .$$

V consists of ℓ pairs of vectors – one for each node – where (u_i,v_i) is the pair for n_i.

where: for $i = 1,2,\ldots,\ell$

$$j = 1,2,\ldots,t$$

R. E. Miller

$$(u_i)_j \begin{cases} -T_j & \text{if } j \in I(n_i) \\ \\ 0 & \text{otherwise.} \end{cases}$$

$$(v_i)_j = \begin{cases} -W_j & \text{if } j \in I(n_i) \cap \overline{O(n_i)} \\ U_j & \text{if } j \in O(n_i) \cap \overline{I(n_i)} \\ U_j - W_j & \text{if } j \in I(n_i) \cap O(n_i) \\ 0 & \text{otherwise} \end{cases}$$

Proposition: $\mathcal{V}(G)$ is a vector replacement system.

Thus any computation graph can be represented by a vector replacement system. A sequence of node firings in the computation graph corresponds to a path in the reachability set of $\mathcal{V}$.

A simple example of this correspondence is shown below:

$$d = (2,0,1)$$
$$V = \{(u_1, v_1), (u_2, v_2)\}$$
$$= \{((-2,0,-1),(-1,1,-1)),$$
$$((0,-1,0),(0,-1,1))\}.$$

For example, $u_1 = (-2,0,-1)$ results from $T_1 = 2$, index $2 \notin I(n_1)$ and $T_3 = 1$; and $v_1 = (-1,1,-1)$ is defined as follows:

index $1 \in I(n_1) \cap O(n_1)$ so $(v_1)_1 = U_1 - W_1 = 1-2 = -1$

index $2 \in O(n_1) \cap \overline{I(n_1)}$ so $(v_1)_2 = U_2 = 1$

and index $3 \in I(n_1) \cap \overline{O(n_1)}$ so $(v_1)_3 = -W_3 = -1$.

Although this construction is very direct, giving exactly one vector replacement system for any computation graph, it is not a one-one onto relationship. Several computation graphs can result in the same vector replacement system. For our example, the following graph G' has the property that $\mathcal{V}(G') = \mathcal{V}(G)$.

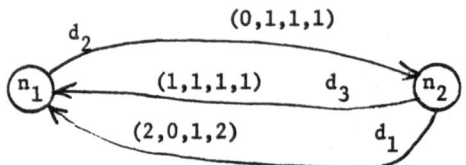

Also, the class of vector addition systems so obtained are quite restrictive.

A tighter relationship between these two models is obtained by further restrictions. Thus, we can state that:

Proposition: The class of all productive irreflexive computation graphs is isomorphic to the class of vector replacement systems in which V is constrained as follows:

For each coordinate j there is exactly one $i(j)$ with $(v_{i(j)})_j < 0$ and exactly one $k(j)$ with $(v_{k(j)})_j > 0$.

Here a computation graph is called productive if for each edge d_i, $U_i > 0$ and $W_i > 0$; and irreflexive if there are no self loops in the graph.

This result states that the set of V vectors has, in each coordinate exactly two nonzero entries, one positive and one negative. These correspond to the two ends of each edge in the graph.

A further restriction of the computation graphs to the case that for each d_i, $W_i = T_i$ results in providing an isomorphism to vector addition systems.

Through this relationship the queue lengths in the computation graph are seen to be represented by coordinate values in the reachability sets. However, termination properties of the computation graph are not as

directly related to the reachability set problem. Note however, that
if the computation graph terminates then the reachability set is finite.
Thus, since an algorithm for termination exists, one can determine if
the computation graph terminates. If it does, then the reachability
set is finite and can be completely determined. However, nontermination
does not imply automatically that the reachability set is infinite since
the computation graph might have bounded queues but not terminate. This
gives some insight, then, into how the results between these two formu-
lations are intimately, but not exactly, related.

Semaphores and Producer-Consumer Systems

One of the important concepts in synchronization of cooperating
processes was put forth by Dijkstra [3] in his notion of semaphores.

A semaphore s may be considered to be a non-negative integer
valued variable upon which only the two following types of primitive
operations may be performed.

$$P(s) \qquad L: \text{if } s < 1 \text{ go to } L \text{ else } s - 1$$

$$V(s) \qquad s \longleftarrow s+1.$$

The P(s) and V(s) operations are each assumed to be indivisible.
That is, once started no other P or V operation can effect s before
the first is completed.

One common use for semaphores is to control the starting and stop-
ping of communicating processes. A simple such system we call an
unshared producer-consumer system. It consists of:

(i)   a finite set $B = p_1, \ldots, p_\ell$ of processes

(ii)   a finite set $S = s_1, \ldots, s_t$ of semaphores

(iii)   a function $\alpha: S \to B \times B$, associating each semaphore to a pair
          of processes.

R. E. Miller

(iv)   Three functions:

μ: S → N   (initial value)

π: S →   (# of P's)

ν: S → N   (# of V's) .

This definition can be viewed as the following connection between any

two processes p_i and p_j.

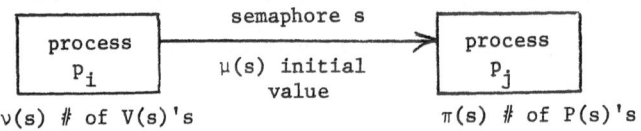

Here process P_i (the producer) is assumed to end with ν(s) V(s)

operations, which increases the value of s by ν(s). Process p_j

(the consumer) is assumed to start with π(s) P(s) operations. Thus a

loose sequencing is maintained between p_i and p_j; i.e., p_j cannot start

without s being sufficiently large (π(s) or more) and this may depend

upon process p_i having completed some previous executions.

    A more general <u>producer-consumer system</u> can be defined consisting of:

(i)   a finite set B = $\{p_1,...,p_\ell\}$ of processes,

(ii)   a finite set S = $\{s_1,...,s_t\}$ of semaphores;

(iii)   Three functions:

μ: S → N   initial value

π': S×B → N   (# of P(s) at process)

ν': S×B → N   (# of V(s) at process)

    This type of producer-consumer system does not restrict a semaphore

to being associated with only one pair of processes.

R. E. Miller

Computation Graphs and
Unshared Producer-Consumer Systems

By a straightforward construction we can show:

Proposition: Any unshared producer-consumer system can be represented by a computation graph.

The computations performed by the computation graph can be seen to simulate those of the unshared producer-consumer system. Here it is important that the producer-consumer system is "unshared." That is, that each semaphore is associated only with a single pair of processes. This can be seen by noting that a starting of an operation in the computation graph removes a certain fixed number of operands, in one chunk, from input queues, and on completion places a number of results in a chunk in output queues. This differs from the producer-consumer operation in that P's and V's successively decrease or increase the values only by one at a time. Thus, iteration by P's or V's from several processes on a single semaphore could cause some different or unexpected operation. For the unshared situation this cannot occur.

Producer-Consumer Systems and
Generalized Petri Nets

A generalized Petri net is similar to a Petri net except that firing of transitions may remove or add more than one token at a time to a place. Formally a

generalized Petri net $P = (\Pi, \Sigma, R, M_0, \Delta_I, \Delta_0)$

where Π is a finite set of places, Σ is a finite set of transitions $R \subseteq (\Pi \times \Sigma) \ (\Sigma \times \Pi)$ describing the edges, $M_0 : \Pi \to N$ is the initial marking and the two functions

$$\Delta_I : \Pi \times \Sigma \to N$$

and $\quad \Delta_0 : \Sigma \times \Pi \to N$

R. E. Miller

with value 0 if $\Pi \times \Sigma$ or $\Sigma \times \Pi$ is not in R.

It is the Δ_I and Δ_0 functions that determine for any place and transition pair, how many tokens are removed or added to the place due to the transition firing.

Now we can obtain the following relationship:

Proposition: Any producer-consumer system can be represented by a generalized Petri net, and conversely.

The representation associates items as follows:

S		P
processes B	$\sim$	transitions Σ
semaphores S	$\sim$	places Π
$\mu(s_j)$	$\sim$	$M_0(\pi_j)$
$\pi'(s_i,p_j)$	$\sim$	$\Delta_I(\pi_i,\sigma_j)$
$\nu'(s_i,p_j)$	$\sim$	$\Delta_0(\sigma_j,\pi_i)$

A simple example illustrating this relationship is the well-known mutual exclusion problem. It is shown both as a producer-consumer system and as a generalized Petri net below.

Mutual Exclusion Problem

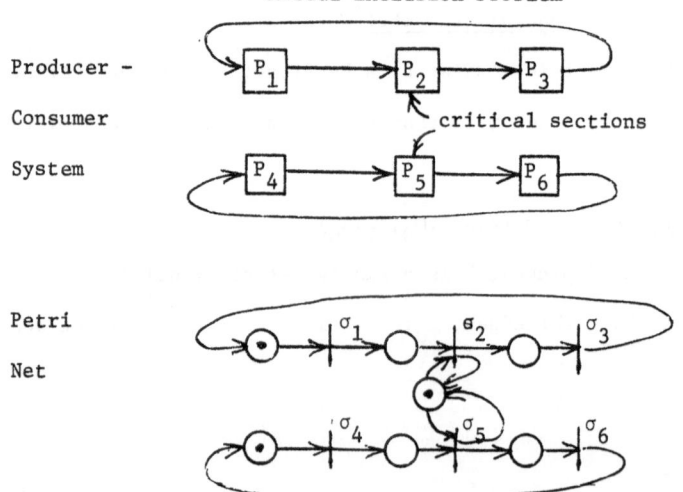

R. E. Miller

As mentioned earlier, independence of the P and V operations, changing a single semaphore by only one, can create a problem when one considers whether the "behavior" of the two systems are equivalent or not. Another example illustrates this simply.

Consider the producer-consumer system with three processes and three semaphores with

$$P_1: \begin{array}{l} P(s_3) \\ V(s_1) \\ V(s_2) \end{array} \qquad P_2: \begin{array}{l} P(s_1) \\ P(s_2) \\ V(s_3) \end{array} \qquad P_3: \begin{array}{l} P(s_2) \\ P(s_1) \\ V(s_3) \end{array}$$

This translates to the generalized Petri net

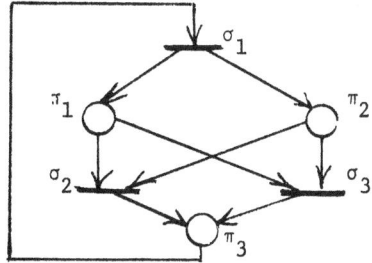

Here the producer-consumer system can "deadlock" if after p_1 operates process p_2 performs $P(s_1)$ and process p_3 performs $P(s_2)$. The Petri net does not deadlock since either σ_2 or σ_3 would fire, simultaneously removing a token from π_1 and π_2. Thus, the conflict situations in Petri nets, or the possible deadlock situations in producer-consumer systems are not modelled appropriately in this relationship between the two models. In fact, more complex examples show that even by reordering the P's at the start of processes cannot always circumvent this sort of difficulty.

R. E. Miller

Generalized Petri Nets and Vector Addition and Vector Replacement Systems

To show relationships between these models we define two new ideas.
Two transitions $\sigma \neq \sigma'$ of a generalized Petri net are called equivalent
transitions if and only if, for all $\pi \in \Pi$

$$\Delta_I(\pi,\sigma) = \Delta_I(\pi,\sigma') \text{ and}$$

$$\Delta_0(\sigma,\pi) = \Delta_0(\sigma',\pi).$$

A generalized Petri net is called irreflexive if and only if there
does not exist any $\pi \in \Pi$ and $\sigma \in \Sigma$ such that $\Delta_I(\pi,\sigma) > 0$ and $\Delta_0(\sigma,\pi) > 0$.
That is there are no self-loops.

Proposition: There is an isomorphism between vector replacement systems
and generalized Petri nets without equivalent transitions.

Proposition: There is an isomorphism between vector addition systems
and irreflexive generalized Petri nets without equivalent transitions.

In both of these isomorphisms there is a direct correspondence
between the firing sequences in the Petri nets and the reachable paths
in the vector systems. These lead to interesting connections of the
model properties.

References

[1]   J. L. Baer, "A Survey of Some Theoretical Aspects of Multi-
        processing," ACM Computing Surveys, 5, 1, 31–80, March, 1973.

[2]   F. Commoner, A. W. Holt, S. Even, A. Pneuli, "Marked Directed
        Graphs," Journal of Computer and Systems Science, Vol. 5,
        October 1971, pp. 511-23.

[3]   E. W. Dijkstra, "Co-operating Sequential Processes," in
        Programming Languages, F. Genuys, Editor, New York, Academic
        Press, 1968, pp. 43-112.

[4]   M. Hack, "Decision Problems for Petri Nets and Vector Addition
        Systems," MIT Project MAC, Computation Structures Group Memo 95,
        March 1974.

R. E. Miller

[5]   R. M. Keller, "Vector Replacement Systems: A Formalism for
        Modeling Asynchronous Systems," Princeton University,
        E. E. Technical Report No. 117, December 1972.

[6]   R. M. Karp and R. E. Miller, "Properties of a Model for
        Parallel Computation: Determinacy, Termination, Queueing,"
        SIAM Journal of Applied Math., 1390-1411, November 1966.

[7]   R. M. Karp and R. E. Miller, "Parallel Program Schemata,"
        Journal of Computer and System Sciences, 3, 2, 147-195, May 1969.

[8]   R. E. Miller, "A Comparison of Some Theoretical Models of
        Parallel Computation," IEEE Transactions on Computer, C-22, 8,
        710-717, August 1973.

[9]   R. E. Miller, "Some Relationships Between Various Models of
        Parallelism and Synchronization." IBM Research Report
        RC-5074, October 1974.

[10]   W. L. Miranker, "A Survey of Parallelism in Numerical Analysis,"
        SIAM Review, 13, 524-547, October 1971.

[11]   C. A. Petri, "Communication with Automata," Supplement 1 to
        Technical Report RADC-TR-65-337, 1, Griffiss Air Force Base,
        New York, 1966 [Translated from German "Kommunikation mit
        Automaten," University of Bonn, 1962.]

[12]   M. Yoeli, "Petri Nets and Asynchronous Control Networks,"
        Applied Analysis and Computer Science Research Report
        CS-73-07, University of Waterloo, Waterloo, Ontario, Canada,
        April 1973.

CENTRO INTERNAZIONALE MATEMATICO ESTIVO
(C.I.M.E.)

THEORY OF AUTOMATA

D.E. MULLER

Corso tenuto a Bressanone dal 9 al 17 giugno 1975

THEORY OF AUTOMATA

David E. Muller[*]

1.   The title "theory of automata" has become somewhat passe because
the problems of interest to people working in this area have changed,
and therefore the subject has split into a number of different
disciplines. Each of these disciplines is called by its own name, but
many of the results and techniques developed in the theory of automata
have turned out to be useful. Our object is to look at a few of these
results and techniques and see what sorts of approaches have been
applied to problems involving automata.

Historically, when people first began to design computing machines,
it became clear that the interaction of the various units was
sufficiently complex so that systematic methods for organizing the design
were necessary. The different parts of the machine were thought of as
units and ways were worked out for specifying the design of a unit in
terms of the organization of the machine as a whole. Also, the design
of a unit had to be systematized.

The methods employed were algebraic, and consequently much of the
theory of automata has retained an algebraic flavor. Thus, we design
circuits using switching algebra, we discuss homomorphisms among
machines, and we use semigroups in the study of cascade decomposition.

[*]Coordinated Science Laboratory and Department of Mathematics,
University of Illinois at Urbana-Champaign. This work was supported in
part by the Joint Services Electronics Program (U. S. Army, U. S. Navy,
and U. S. Air Force) under Contract DAAB-07-72-C-0259.

D. E. Muller

Also, in the early days many logicians became interested in
problems concerning automata. A possible reason for this interest was
the success of A. M. Turing in 1936 in applying the notion of a machine
to characterize an effective computation and thus to obtain some of the
original results concerning unsolvability. In any case, many logicians
turned their attention to the study of automata, and as a consequence
the nature of the subject and types of questions which have been asked
and answered were profoundly influenced the bent of these early workers.

One person who stimulated many others to study automata was
John Von Neumann. He was interested both in designing computers more
intelligently and in designing more "intelligent" computers. The reason
for applying the word "intelligent" to computers here is that Von Neumann
thought of the computer as a model for the brain. While the intervening
25 years since Von Neumann's work have not seen great advances in
forming a computer model of the action of the brain, nevertheless, this
idea spurred much of the original investigation of machine theory and
is still being actively pursued today.

In the model used by Von Neumann the basic organ was a device with
several input lines and one output line. The lines were capable of
carrying signals which were either 0 or 1. The signal carried by the
output line was determined by the combination of signals on the input
lines after a fixed time delay τ. It was thought necessary to use
synchronizing signals to ensure proper timing in an actual machine, and
in the mathematical model such synchronization was achieved by allowing
a fixed and constant time delay between the occurrences of the
combination of input signals and the resulting output signal.

A network composed of such organs in suitable variety and suitably

connected could do "anything." That is, if the network was able to
control devices capable of writing on and reading from a potentially
infinite roll of paper tape then it could be made to carry out any
computation which could be effectively described. The machine of Turing,
or for that matter any other computing device could be designed using
the elements of Von Neumann.

The elements in the brain were thought to be (1) slow compared with
electronic speeds and (2) unreliable compared to manufactured components,
but they were much smaller than the devices available in those early
years and their organization was much more complicated. Regular signals
travelling through the nervous system were thought by Von Neumann and
others to be synchronizing signals. But, there remained the problem of
reliability. Was it possible for even the rather unreliable elements
found in the nervous system to produce reliable results? Von Neumann
proved that it was. As long as the elements were not totally unreliable,
he showed that the network as a whole could be made to function with
arbitrarily good, although not perfect, reliability.

To produce a reliable automaton using unreliable elements,
Von Neumann used an approach which is extravagant in its use of
equipment although versions of it still find favor today. First, a
design is worked out on the assumption that each organ works perfectly,
never failing to produce the correct output signal. From this original
design one obtains the final one by a substitution process. Each line
of the original design is replaced by a bundle containing some large
number n of actual lines in the final one. Let us assume that n is a
multiple of 3. Each organ of the original design is correspondingly
replaced by n similar organs which are connected in like manner to the

D. E. Muller

ı lines of the bundles. Finally, errors occurring in a few of the n organs are partially corrected in the bundles by placing within each of them n/3 "voting" organs which, when functioning properly, produce an output which agrees with the majority of the three inputs. The n/3 outputs of the "voting" organs are each split into three lines and the resulting n lines are randomly scrambled. Von Neumann proved that by choosing sufficiently large n, the probability of failure of the entire network could be made arbitrarily small, assuming the component organs were not totally unreliable. In this case, an organ with a reliability of 50% must be considered totally unreliable, since this means that the output signal is totally uncorrelated with the input combination; the output is equally likely to be right or wrong.

Since in the final design single signals were replaced by sets of n signals it was necessary to reinterpret the action of the final network made reliable by replication. Problems of its communication with the outside world remained as sources of potential unreliability which could not be solved by Von Neumann's technique.

2.   There is a close connection between the theory of automata and certain basic concepts of algebra, and by regarding automata from an algebraic point of view it has been possible to apply the theory to many things. It has been applied to the study of formal languages of the type used in computer programming, and to the mathematically exact description of algorithms, processes, and other formal systems. In particular, it has been used to obtain decidability results about statements in formal systems. Another recent example of such an application of the theory of automata is to the simpler Burnside's problem--"Is there an infinite,

finitely generated, periodic group?" A short construction using
automata obtains the result which had previously been discovered by
other methods. The answer is that for each prime p, there is an
infinite p-group with two generators.

Thus, we see that the subject we are investigating has found wide-
spread application outside the realm of computer design.

Let us now get down to specifics. An automaton is an idealized
machine, and therefore it will be thought of as a device which receives
signals from the outside world through a channel called its <u>input</u>. It
carries out a transformation of what it has received and presents the
results back to the world through a channel called its <u>output</u>.
Generally, the input and output channels are constructed by using
several lines with binary (or two valued) signals. For example, if
three binary lines are used in a channel one can obtain eight combinations of signals. The set of combinations actually used with a channel
is called the <u>alphabet</u> associated with that channel. In describing an
automaton, let us represent the input and output alphabets by the
letters Ξ and Γ respectively.

To complete the description of the automaton, let S represent the
set of possible internal conditions or <u>states</u>. The state of the
automaton is characterized by the combination of signals appearing on
the lines internal to the automaton. Thus, S may also be regarded as
an alphabet. Now, according to the assumption that signals are
synchronized, we may treat time as a discrete variable and assume that
all changes must occur at integral times: $0, 1, 2, \ldots$, and that the state
of the automaton at time 0 is given. Then, the state at each time $t+1$
and the output at time t are uniquely determined by the input and state

that exist at time t. The two functions which describe the next state

and output are called the <u>transition</u> and <u>output</u> functions and will be

written M and N respectively.

To summarize, we see that an automaton can be described by

specifying the quantities Ξ, Γ, S, M, and N, where the first three

represent sets, called alphabets and the last two are functions:

$M : S \times \Xi \rightarrow S$, and $N : S \times \Xi \rightarrow \Gamma$. In other words, M maps each pair (s,a),

where $s \in S$ and $a \in \Xi$, to a "next" state M(s,a), abbreviated sa and also

in S. Similarly, N maps each such pair (s,a) to an output N(s,a) in Γ.

If we specify the state s(0) in which the automaton is placed at

time 0 and also specify the inputs a(0),a(1),...,a(r) which occur at

times 0,1,...,r, then we can uniquely determine the states s(1),s(2),...,

s(r+1) which occur at times 1,2,...,r+1 respectively. This is done by

taking s(t+1) = M(s(t),a(t)) = s(t)a(t) for t = 0,1,...,r, recursively.

Similarly, we can determine the sequence b(0),b(1),...,b(r) of outputs

at times 0,1,...,r respectively by taking b(t) = N(s(t),a(t)) for

t = 0,1,...,r.

In actual automata the alphabets must be finite, and hence the

functions M and N can always be described. If no simpler rule is

available, they may be tabulated and their calculation carried out by

table look-up.

3.   While an automaton is defined by specifying the five quantities Ξ,

Γ, S, M, and N, we may choose to ignore the output function N and its

associated alphabet Γ, concentrating just on the alphabets Ξ and S and

the transition function M. Such a triple Ξ, S, M is called a <u>monadic</u>

algebra or <u>partial automaton</u>. To understand the reason that the triple

D. E. Muller

Ξ, S, M can be regarded as an algebra, we note that the mapping

$M : S \times \Xi \to S$ may be restricted by fixing an element $a \in \Xi$ and writing

$M_a : S \to S$. For each $s \in S$, we define $M_a(s)$ as $M(s,a)$. There is thus a

mapping M_a for each $a \in \Xi$. Here, as before, we may abbreviate $M_a(s)$ as

sa, and the function M_a may be thought of as operating upon the state s

to produce the new state sa. The elements of Ξ designate operations of

the algebra which are all unary in the sense that a single state s is

needed to produce the result sa.

A monadic algebra can be represented as a directed graph with labels

on the edges corresponding to the operation they represent. For example,

let $\Xi = \{0,1\}$, $S = \{s_0, s_1, s_2\}$ and let M be given by the table of

Figure 1a. Then, the directed graph of the monadic algebra is shown in

Figure 1b.

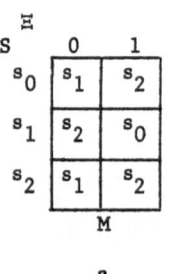

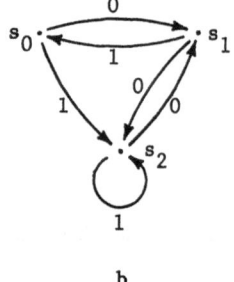

a   b

Figure 1

If s_0 is a state in S, then we define the subset $M(s_0)$ <u>generated</u>

from s_0 as the intersection of all subsets S' of S such that

    (i) $s_0 \in S'$,

    (ii) whenever $s \in S'$ and $a \in \Xi$, then $sa \in S'$.

Intuitively $M(s_0)$ consists of s_0 and all states "reachable" from

D. E. Muller

s_0 by the application of a finite sequence of operations of the algebra
to s_0. In the language of automata, $M(s_0)$ is the set consisting of all
states to which the automaton may be brought by some sequence of inputs
if it is started in state s_0.

Exercise 1: Prove that the formal definition of $M(s_0)$ is valid by
showing that there exists at least one set satisfying the conditions
required of S'. Hint: Show that S itself is such a set.

Exercise 2: Prove that the set $M(s_0)$ satisfies both conditions (i)
and (ii) for S'.

Exercise 3: Prove that $M(s_0)$ has no proper subset satisfying these
two conditions.

Exercise 4: Let $\Xi = \{0,1\}$, $S = \{s_0,s_1,s_2\}$, and let M be defined by
the following table. Show that $M(s_0) = \{s_0,s_2\}$, $M(s_1) = \{s_0,s_1,s_2\}$, and
$M(s_2) = \{s_0,s_2\}$.

	0	1
s_0	s_2	s_2
s_1	s_2	s_0
s_2	s_0	s_2

A monadic algebra with a distinguished <u>initial state</u> or <u>generator</u>
is one in which a certain state s_0 is specially designated and is such
that the entire set S is generated from it. In other words, it must be
such that $M(s_0) = S$. We describe such an algebra by listing Ξ, S, M,
and s_0. Note that in Figure 1, if s_0 is chosen as the distinguished
initial state then the relation $M(s_0) = S$ is satisfied.

Although the notion of a monadic algebra was introduced by first
describing an associated automaton and hence it is natural to think of

the set of states as finite, the finiteness restriction may be relaxed as long as we are merely discussing abstract algebras. As an example of an infinite monadic algebra, let Ξ consist of a single letter, say $\Xi = \{'\}$, and let S consist of the natural numbers 0,1,2,... . Then define the transition function by the rule that for each state $s \in S$, let $s' = s+1$. In this example it is possible to choose only 0 as the distinguished initial state. Naturally, since S is infinite it is impossible to give a tabulation of the transition function or a directed graph corresponding to the algebra.

The following theorems are very useful if we wish to carry out formal proofs of properties of monadic algebras having distinguished initial states. One can construct parallel theorems about any algebra with a distinguished set of generators.

Theorem 1: (Induction) For any monadic algebra (Ξ,S,M) with a distinguished initial state s_0, let P(s) be a proposition when $s \in S$. Suppose that

(i) $P(s_0)$ holds, and

(ii) whenever P(s) holds and a $\in \Xi$, P(sa) holds.

Then P(s) must hold for all $s \in S$.

Proof: Let S' be the subset of S consisting of all s such that P(s) holds. Then $s_0 \in S'$, and if $s \in S'$ then sa $\in S'$ by (i) and (ii) of the hypothesis. In exercise 3 it was shown that $M(s_0) = S$ includes no proper subsets satisfying these two conditions, so S' = S. □

Theorem 2: For any monadic algebra (Ξ,S,M) with a distinguished initial state s_0, if $s \in S$ but s is not s_0, then s can be expressed in the form s'a for some s' $\in S$ and a $\in \Xi$.

Proof: Let P(s) be the proposition: either s = s_0 or s may be expressed in the form s'a. Then, referring to theorem 1, we see that

(i) $P(s_0)$, and (ii) if $P(s)$ and $a \in \Xi$, then $P(sa)$. We note that in (ii)

the conclusion $P(sa)$ always holds even if it could happen that $P(s)$ did

not. Naturally, this fact does not invalidate (ii). Thus $P(s)$ must hold

for all $s \in S$, by theorem 1. This proves theorem 2. □

The two theorems just obtained are not usually presented in a formal

way when discussing automata. The reason we have done so is that in

order to see the parallel between conventional automata and various

generalizations such as tree automata it is convenient to write every-

thing down so that the differences and similarities between the theories

are clear.

To illustrate this technique of generalization we shall briefly

introduce the theory of tree automata. In our treatment of this theory,

we shall use the notion of a groupoid to replace that of a monadic

algebra while keeping the theory as similar as possible in all other

respects.

A groupoid is an algebra with a single binary operation. There are

no specialized properties such as associativity orcommutativity required

of all groupoids and there is no alphabet such as Ξ associated with a

groupoid. Briefly therefore, a groupoid consists of a set S and a

mapping $M : S \times S \to S$. To emphasize the parallel between the two systems,

we shall call S the set of states and M the transition function. For

convenience and to avoid confusion, we shall abbreviate the binary

operation $M(s,s')$ as $s \gamma s'$.

As in the case of monadic algebras, define the subset $M(s_0)$ of

states generated from s_0 as the intersection of all subsets S' of S such

that

D. E. Muller

(i) $s_0 \in S'$, and

(ii) whenever s, s' $\in S'$, then s γ s' $\in S'$.

Results similar to those of exercises 1, 2, and 3 may be proved in exactly the same way.

A groupoid (S,M) with a distinguished <u>generator</u> s_0 must satisfy the relation $M(s_0) = S$, just as in the case of a monadic algebra.

<u>Exercise 5</u>: State and prove the theorems analogous to theorems 1 and 2 for the case of a groupoid. We shall call these theorems 1' and 2'.

We now return to our original treatment of monadic algebras and define a <u>free monadic algebra</u> on an alphabet Ξ as a monadic algebra with input alphabet Ξ and a distinguished initial state s_0 which has the following two properties.

(1) For all $s \in S$ and $a \in \Xi$, sa $\neq s_0$, and

(2) for all s, s' $\in S$ and a, a' $\in \Xi$, if sa = s'a', then s = s' and a = a'.

Although it is customary to define free algebras in a different way and involving the notion of a homomorphism, we shall see that this definition turns out to be equivalent to the usual one.

<u>Lemma 1</u>: There exists a free monadic algebra on any alphabet Ξ.

Proof: Let us assume that we are given the alphabet Ξ, then we proceed to construct a monadic algebra we shall call $\mathfrak{F}(\Xi)$ using Ξ as its alphabet by the following process. Let Λ be a symbol which is not a member of Ξ and take Λ as the initial state of $\mathfrak{F}(\Xi)$. Define the transition function M by the general rule that if t is any quantity, then for $a \in \Xi$, take M(t,a) to be the ordered pair (t,a). Finally, let the state set T of $\mathfrak{F}(\Xi)$ consist of all elements generated by M from Λ.

D. E. Muller

It is clear from the definition that $\mathcal{F}(\Xi)$ as defined here is a monadic algebra with the distinguished initial state Λ. To see that it is free in the sense we defined previously, we note first that Λ is merely a symbol and hence cannot be an ordered pair of the form ta. Hence (1) is satisfied. Second, let us assume that ta = t'a'. Thus, (t,a) = (t',a'), and since two ordered pairs are equal iff their first and second members are respectively equal, we have t = t' and a = a'. Hence, (2) is satisfied and $\mathcal{F}(\Xi)$ is free. □

Although we have used a particular method for constructing $\mathcal{F}(\Xi)$ we shall now show that $\mathcal{F}(\Xi)$ is the only free monadic algebra on Ξ up to isomorphism. We also will see that the elements of T may be regarded as finite sequences of letters from Ξ. This result is important because we recall that such finite sequences were applied at the input channel of an automaton.

We begin by defining a mapping which will be important in our proof. Let $\mathcal{C}$ be a monadic algebra (Ξ, S, M_1) and, as before, let $\mathcal{F}(\Xi)$ be the free monadic algebra (Ξ, T, M) of lemma 1. Then, define the mapping $M_{\mathcal{C}} : S \times T \to S$ by the following two rules, where st is written as an abbreviation for $\mathcal{M}_{\mathcal{C}}(s,t)$.

    (i) For all $s \in S$, $s\Lambda = s$.

    (ii) For all $s \in S$, $t \in T$, $a \in \Xi$, s(ta) = (st)a.

If rule (ii) were written without using our abbreviations it would appear as $\mathcal{M}_{\mathcal{C}}(s, M(t,a)) = M_1\,\mathcal{M}_{\mathcal{C}}(s,t), a)$.

It is necessary to prove that $\mathcal{M}_{\mathcal{C}}$ is well defined by the above definition which is of an inductive nature. That is, for each $s \in S$ and $t \in T$ there is exactly one product st. To carry out this proof we imagine that s is fixed.

D. E. Muller

First, let t be Λ. Then, rule (i) above may be used for defining
the product st. By rule (1) in the definition of a free monadic algebra,
we see that t cannot be of the form t'a. Thus, rule (ii) above cannot
be applied, so the product st is uniquely given by (i).

Second, we use induction (theorem 1) as applied to $\mathfrak{F}(\Xi)$. We make
the inductive assumption that for given $t \in T$ the product st is
unambiguously defined, i.e. is unique. We just saw that this assumption
is true when t is Λ. Now consider the product s(ta). By (ii) above,
we see that one value for this product is (st)a. Since rule (1) tells
us that ta cannot be Λ, we see that (i) is not applicable in this case.
Thus, for s(ta) to be more than single valued it would be necessary to
apply rule (ii) a second time. This would require that ta = t'a' so
that s(ta) = s(t'a') = (st')a'. But rule (2) in the definition of a
free monadic algebra tells us that ta = t'a' implies that t = t' and
a = a', so (st')a' = (st)a and the two values we assumed for s(ta) must
be the same and it is unambiguously defined. This completes the
inductive step and the function $\mathfrak{M}_{\alpha}$ is well defined.

The function $\mathfrak{M}_{\alpha}$ may be regarded as an extension of the transition
function M_1 of the monadic algebra α. Let s be the state of α at some
time t, where α is now regarded as a partial automaton. Then, we have
seen that sa is the state at time t+1 if the input a is applied at time
t. Now, we shall see that each element t of T corresponds to a sequence
$a_1, a_2 \ldots a_n$ of letters from Ξ. If these letters are applied to the input
successively during the present and next n-1 time units, then the state
at time t+n will be just st. These facts will be stated more formally
in a later theorem.

In the definition of $\mathfrak{M}_{\alpha}$ the two monadic algebras $\mathfrak{F}(\Xi)$ and α were both involved. Taking $\mathcal{Q}$ to also be $\mathcal{F}(\Xi)$, we define $m_{\mathcal{F}(\Xi)} : T \times T \to T$. For any two elements t, t' $\in$ T, we abbreviate $m_{\mathcal{F}(\Xi)}(t,t')$ as tt'. Then, we can prove the following.

Theorem 3: Let $\mathcal{Q}$ be any monadic algebra (Ξ,S,M_1) and let t, t' be any two elements of T, the state set of $\mathcal{F}(\Xi)$, then (st)t' = s(tt').

Proof: It is clearly true if t' = Λ by property (i) in the definitions of $m_{\mathcal{Q}}$ and $m_{\mathcal{F}(\Xi)}$. Using induction, assume it holds for given t' $\in$ T. Then, for a $\in \Xi$, we have (st)(t'a) = ((st)t')a = (s(tt'))a = s((tt')a) = s(t(t'a)), using property (ii) in the definitions of $m_{\mathcal{Q}}$ and $m_{\mathcal{F}(\Xi)}$. Hence, it holds for all t' $\in$ T. □

Theorem 4: In $\mathcal{F}(\Xi)$, for all t $\in$ T, Λt = t.
Proof: It is true when we take t = Λ by (i) in the definition of $m_{\mathcal{F}(\Xi)}$. Assuming inductively that it holds for given t and taking a $\in \Xi$, we have Λ(ta) = (Λt)a = ta. □

This last theorem makes it reasonable to introduce the abbreviation of Λa as a. Although Λa is a member of T and a is in Ξ, no confusion will result from this abbreviation because we have adopted the same notation for M_1 and for $m_{\mathcal{Q}}$. Since s(Λa) = (sΛ)a = sa, the new notation may always be used. Algebraically we have identified the set Ξ with the subset $\{\Lambda a | a \in \Xi\}$ of T, and we shall henceforth regard Ξ as a subset of T.

Exercise 6: Using the monadic algebra of figure 1, construct the product s(10) for each state s $\in$ S. Hence, show that theorem 3 holds for the case $(s_0 0)(10) = s_0(0(10))$.

Results analogous to theorems 3 and 4 are not convenient to state in the groupoid case. It is possible, however, to define a free groupoid with a distinguished generator s_0 as a groupoid having the following two properties.

    (1) For all $s,s' \in S$, $s \curlyvee s' \neq s_0$, and

    (2) for all $s,s',s'',s''' \in S$, if $s \curlyvee s' = s'' \curlyvee s'''$, then $s = s''$ and

       $s' = s'''$.

The analog of lemma 1 may then be obtained in which it is shown that a

free groupoid $\mathfrak{F}_1$ exists.

    Exercise 7: State and prove lemma 1', the analog of lemma 1 for

groupoids. Define the transition function of $\mathfrak{F}_1$ by the rule that $s \curlyvee s'$

is the ordered pair (s,s').

4.   Returning to the case of monadic algebras, define a <u>homomorphism</u>

from a monadic algebra (Ξ,S,M) to another monadic algebra (Ξ,S',M') on

the same input alphabet Ξ as a mapping $h : S \to S'$ such that for all $s \in S$

and $a \in \Xi$, $h(sa) = h(s)a$. In this definition we have used our standard

abbreviation in an unambiguous way. Without the abbreviation the

equation $h(sa) = h(s)a$ becomes $h(M(s,a)) = M'(h(s),a)$.

    Theorem 5: If h is a homomorphism, then for all $s \in S$ and $t \in T$,

$h(st) = h(s)t$.

    Proof: The theorem is clearly true when $t = \Lambda$. Using induction

on $\mathfrak{F}(\Xi)$, assume inductively that it holds for given t. Then $h(s(ta)) =$

$h((st)a) = h(st)a = (h(s)t)a = h(s)(ta)$. This completes the inductive

step and the result is proved. $\square$

    Theorem 6: Let (Ξ,S,M) be a monadic algebra with a distinguished

initial state s_0. Then there is a homomorphism h from $\mathfrak{F}(\Xi)$ to (Ξ,S,M)

such that $h(\Lambda) = s_0$. This homomorphism is unique, onto, and given by

the rule $h(t) = s_0 t$ for all $t \in T$.

    Proof: Let us define the mapping $h : T \to S$ by the rule that

$h(t) = s_0 t$. To prove that this mapping is a homomorphism we observe

that for any $t \in T$ and $a \in \Xi$, we have $h(ta) = s_0(ta) = (s_0 t)a = h(t)a$.

Clearly, $h(\Lambda) = s_0$.

To prove uniqueness, let $h' : T \to S$ be any other homomorphism such

that $h'(\Lambda) = s_0$. By theorems 4 and 5, $h'(t) = h'(\Lambda t) = h'(\Lambda)t = s_0 t$.

Since this holds for arbitrary t, we see that $h' = h$.

To prove that h as given above is onto, we use induction on the

algebra (Ξ, S, M). Note that s_0 is the image under h of some member of T,

namely Λ. Assume inductively that $s \in S$ is the image of $t \in T$. Then sa

is the image of ta since $h(ta) = h(t)a = sa$. This completes the

induction step so therefore each $s \in S$ must be the image of some $t \in T$. □

Corollary: Each state $s \in S$ can be expressed in the form $s_0 t$ for

some $t \in T$.

Exercise 8: If (S,M) and (S',M') are two groupoids, then a

homomorphism from (S,M) to (S',M') is defined as a mapping $h : S \to S'$

such that for all $s_1, s_2 \in S$, the relation $h(s_1 \gamma s_2) = h(s_1) \gamma h(s_2)$ holds.

Now, let (S,M) be any groupoid with a distinguished generator s_0 and let

ϕ be the generator of the free groupoid $\mathfrak{F}_1$ of exercise 7. Prove that

there is a homomorphism from $\mathfrak{F}_1$ to (S,M) such that $h(\phi) = s_0$ and that

this homomorphism is unique and onto. Hint, define h(t) by the rules:

(1) $h(\phi) = s_0$, and (2) $h(t_1 \gamma t_2) = h(t_1) \gamma h(t_2)$.

Theorem 7: Let (Ξ, S, M) be any free monadic algebra with the

distinguished initial state s_0. Then there is an isomorphism from

$\mathfrak{F}(\Xi)$ onto (Ξ, S, M) which maps Λ to s_0.

Proof: By theorem 6, we know there is a homomorphism h from $\mathfrak{F}(\Xi)$

onto (Ξ, S, M) such that $h(\Lambda) = s_0$ and that this homomorphism is unique.

It merely remains, therefore, to show that h is one-to-one.

D. E. Muller

We begin by showing that if $h(t) = s_0$, then $t = \Lambda$. Assume the contrary. By theorem 2, $t = t_1 a$ for some $t_1 \in T$ and $a \in \Xi$. Then, $s_0 = h(t) = h(t_1 a) = h(t_1)a$, which contradicts property (1) of a free monadic algebra. Thus s_0 has just one preimage under h, namely $^*\Lambda$. Now, we assume inductively that a given $s \in S$ has just one preimage, say $t \in T$. For any $a \in \Xi$, clearly ta is a preimage of sa. Let t_2 be a second preimage of sa. Since $t_2 \neq \Lambda$, we may write $t_2 = t'a'$ for some $t' \in T$ and $a' \in \Xi$. Hence, $h(t'a') = h(ta)$, and since $h(t'a') = h(t')a'$ and $h(ta) = h(t)a = sa$, we have $h(t')a' = sa$. Since (Ξ, S, M) is a free monadic algebra, it satisfies property (2) and $h(t') = s$, while $a' = a$. Thus, by our inductive hypothesis $t' = t$ and we obtain $t'a' = ta$. This shows that sa has just one preimage, namely ta. This completes the inductive step and by theorem 1, each state in S has just one preimage under h. Hence h is an isomorphism. $\square$

This result shows that all free monadic algebras on Ξ are isomorphic. We may therefore speak of the free monadic algebra on Ξ, and we shall continue to denote it by $\mathfrak{F}(\Xi)$. A similar result may be proved analogously for groupoids.

We have seen that by letting $\mathcal{Q}$ be $\mathfrak{F}(\Xi)$ in the definition of $\mathcal{M}_{\mathcal{Q}}$, the mapping $\mathcal{M}_{\mathfrak{F}(\Xi)}$ defines a binary operation on T. By theorem 3, applied to this case, we see that the binary operation is associative. Hence, we say that $\mathcal{M}_{\mathfrak{F}(\Xi)}$ defines a semigroup, and we shall denote it by $\mathscr{A}(\mathfrak{F}(\Xi))$. From the definition of $\mathcal{M}_{\mathfrak{F}(\Xi)}$, part (i), we see that in this semigroup $t\Lambda = t$ for all $t \in T$. Also, by theorem 4, we have $\Lambda t = t$. Hence Λ is an identity element in $\mathscr{A}(\mathfrak{F}(\Xi))$. We notice that the elements of $\mathfrak{F}(\Xi)$ and $\mathscr{A}(\mathfrak{F}(\Xi))$ are identical. The difference lies in the fact that $\mathfrak{F}(\Xi)$ is an algebra with a number of unary operations while $\mathscr{A}(\mathfrak{F}(\Xi))$ has a single

D. E. Muller

binary operation.

Although we shall not continue the detailed development, it can be shown by techniques similar to those we have already used that $\mathscr{S}(\mathfrak{F}(\Xi))$ is the <u>free monoid</u> on Ξ. This is the same as the free semigroup to which an identity element Λ has been added. As mentioned earlier, the elements of T may be regarded as the finite sequences or strings of letters from the alphabet Ξ with Λ representing the sequence of length zero. The binary operation of $\mathscr{S}(\mathfrak{F}(\Xi))$ is thus concatenation, so that if $t = a_1 \ldots a_n$ and $t' = a_1' \ldots a_m'$, then $tt' = a_1 \ldots a_n a_1' \ldots a_m'$. Furthermore, the unary operations of $\mathfrak{F}(\Xi)$ may be shown to correspond to the operation of attaching a letter to the right hand end of a string. Thus, if $t = a_1 \ldots a_n$, then $ta = a_1 \ldots a_n a$.

5.   Let $\mathcal{Q} = (\Xi, S, M)$ be a monadic algebra, then in $\mathfrak{F}(\Xi)$ each element $t \in T$ yields a mapping $\mathcal{M}_t : S \to S$ by the rule that the image of each $s \in S$ under $\mathcal{M}_t$ is st. The set of all such mappings is closed under functional composition since for $t, t' \in T$ and $s \in S$ we have $\mathcal{M}_t \mathcal{M}_{t'}(s) = \mathcal{M}_t(st') = st't = \mathcal{M}_{t't}(s)$. Hence the set of all such mappings forms a semigroup called the <u>semigroup of</u> $\mathcal{Q}$ which we shall denote by $\mathscr{S}(\mathcal{Q})$. We note that this semigroup is a semigroup with an identity, called a <u>monoid</u>, since $\mathcal{M}_\Lambda$ is the identity map on S. We note that if S is finite then $\mathscr{S}(\mathcal{Q})$ must be finite also because there are only finitely many mappings from a finite set to itself.

  <u>Exercise 9</u>: If $\mathscr{S}(\mathcal{Q})$ is finite must S be finite also?

  <u>Exercise 10</u>: Give an example in which $\mathscr{S}(\mathcal{Q}_1)$ and $\mathscr{S}(\mathcal{Q}_2)$ are isomorphic yet $\mathcal{Q}_1$ has two states while $\mathcal{Q}_2$ has three states.

Exercise 11: Explain why $\mathscr{A}(\mathfrak{J}(\Xi))$ as defined here is consistent with our original definition of $\mathscr{A}(\mathfrak{J}(\Xi))$.

In the following example, let $\Xi = \{0,1\}$ and $S = \{s_0, s_1, s_2\}$, with M defined by the table

	0	1
s_0	s_1	s_2
s_1	s_2	s_2
s_2	s_1	s_2

We construct the semigroup of this monadic algebra by calculating each mapping $\mathfrak{M}_t$ for strings $t \in T$ until no new mappings can be constructed. This is done in the following table in which each mapping $\mathfrak{M}_t$ is labeled with its string t.

s \\ t	0	1	00	000	01	10	11
s_0	s_1	s_2	s_2	s_1	s_2	s_1	s_2
s_1	s_2	s_2	s_1	s_2	s_2	s_1	s_2
s_2	s_1	s_2	s_2	s_1	s_2	s_1	s_2

The subscripts giving distinct members of the semigroup are $\{\Lambda, 0, 1, 00, 10\}$. It has the following table.

	Λ	0	1	00	10
Λ	Λ	0	1	00	10
0	0	00	1	0	10
1	1	10	1	1	10
00	00	0	1	00	10
10	10	1	1	10	10

D. E. Muller

Theorem 8: If h is a homomorphism from a monadic algebra α = (Ξ,S,M) onto another monadic algebra $\alpha' = (\Xi,S',M')$ then there is a mapping H from $\mathscr{A}(\alpha)$ onto $\mathscr{A}(\alpha')$ such that the image of $\mathcal{M}_t$ of $\mathscr{A}(\alpha)$ is the corresponding element $\mathcal{M}'_t$ of $\mathscr{A}(\alpha')$ and which is a semigroup homomorphism.

Proof: If we attempt to define H by the rule that $H(\mathcal{M}_t) = \mathcal{M}'_t$, then we must show that this definition is unambiguous. Suppose two distinct members t,t' of T yield the same element of $\mathscr{A}(\alpha)$ so that $\mathcal{M}_t = \mathcal{M}_{t'}$. Then we must show that they also yield the same element of $\mathscr{A}(\alpha')$, i.e. that $\mathcal{M}'_t = \mathcal{M}'_{t'}$. Let s' be any state in S'. Since h is "onto" we can find a state $s \in S$ such that $h(s) = s'$. Now, $\mathcal{M}'_t(s') = s't = h(s)t = h(st) = h(\mathcal{M}_t(s)) = h(\mathcal{M}_{t'}(s)) = h(st') = h(s)t' = s't' = \mathcal{M}'_{t'}(s')$. Since s' was chosen arbitrarily, we see that $\mathcal{M}'_t = \mathcal{M}'_{t'}$, so there is a mapping H defined by the rule $H(\mathcal{M}_t) = \mathcal{M}'_t$. This mapping is clearly "onto."

We next must show that it is a semigroup homomorphism. Let $\mathcal{M}_{t(1)}$ and $\mathcal{M}_{t(2)}$ be any two members of $\mathscr{A}(\alpha)$. Then $H(\mathcal{M}_{t(2)}\,\mathcal{M}_{t(1)}) = H(\mathcal{M}_{t(1)t(2)}) = \mathcal{M}'_{t(1)t(2)} = \mathcal{M}'_{t(2)}\,\mathcal{M}'_{t(1)} = H(\mathcal{M}_{t(2)})H(\mathcal{M}_{t(1)})$. Thus H is a semigroup homomorphism from $\mathscr{A}(\alpha)$ onto $\mathscr{A}(\alpha')$ and the proof is complete. □

Exercise 12: Give an example such that H is an isomorphism but h is not. If h is an isomorphism must H be one also?

A monadic algebra $\alpha = (\Xi,S,M)$ with a distinguished initial state s_0 is called a <u>semigroup monadic algebra</u> if whenever $s_0t = s_0t'$ for two elements $t,t' \in T$ then for all $s \in S$, $st = st'$.

Theorem 9: If α is a semigroup monadic algebra there is a one-to-one mapping f from $\mathscr{A}(\alpha)$ onto S such that $f(\mathcal{M}_t) = s_0t$.

Proof: Clearly, for each element $\mathcal{M}_t$ of $\mathscr{A}(\alpha)$ the image

D. E. Muller

$f(\mathcal{M}_t) = s_0 t = \mathcal{M}_t(s_0)$ is unique. Also, we can show that f is one-to-one, for let $f(\mathcal{M}_t) = f(\mathcal{M}_{t'})$. Then $s_0 t = s_0 t'$ and since $\mathcal{C}$ is a semigroup monadic algebra $\mathcal{M}_t(s) = st = st' = \mathcal{M}_{t'}(s)$ for all $s \in S$. Hence $\mathcal{M}_t = \mathcal{M}_{t'}$ and f is one-to-one.

Finally, f is "onto" because by theorem 6 each state in S may be represented in the form $s_0 t$ for some $t \in T$. □

In a semigroup monadic algebra we may define a binary operation over S by letting ss' be just st where t is any string such that $s_0 t = s'$. It makes no difference which string t is used because of theorem 9. The semigroup obtained from this binary operation is isomorphic (actually anti-isomorphic because of our notation) to the semigroup $\mathscr{A}(\mathcal{C})$ of the monadic algebra $\mathcal{C}$ by theorem 9.

Exercise 13: Show that the mapping $f : \mathscr{A}(\mathcal{C}) \to S$ of theorem 9 has the property that $f(\mathcal{M}_t \mathcal{M}_{t'}) = f(\mathcal{M}_{t'}) f(\mathcal{M}_t)$.

Exercise 14: Let h be a homomorphism from one semigroup monadic algebra $\mathcal{C}$ to another $\mathcal{C}'$ and let the initial state s_0 of $\mathcal{C}$ map to the initial state s_0' of $\mathcal{C}'$. Show that h is also a semigroup homomorphism.

Exercise 15: Show that for every monoid with generators which can be placed in correspondence with the elements of Ξ, there is a semigroup monadic algebra $\mathcal{C}$ whose semigroup is isomorphic to the monoid.

If $\mathcal{C} = (\Xi, S, M)$ is a monadic algebra, then an equivalence relation ρ over S will be called a congruence relation if whenever $s \rho s'$ i.e. the two states s and s' are related by ρ, then for all $a \in \Xi$, $sa \rho s'a$. Sometimes a congruence relation on a monadic algebra is called a right invariance relation to distinguish it from a semigroup congruence in case $\mathcal{C}$ happens to be a semigroup monadic algebra. We shall, however, specify a semigroup congruence when possibility of confusion exists.

D. E. Muller

The properties of congruence relations on monadic algebras are similar to the properties of congruence relations on other algebras so we shall state the more elementary ones without proof.

If ρ is a congruence relation as described above, then we can use it to construct <u>congruence classes</u> which partition S. Let $[s]$ denote the congruence class containing $s \in S$. The <u>quotient algebra</u> $\mathcal{Q}/\rho$ is one whose states are the ρ congruence classes of $\mathcal{Q}$. In it the transition function is defined by the rule that $[s]a$ is just $[sa]$. One can show that this definition is unambiguous. Furthermore, there is a "natural" homomorphism h from $\mathcal{Q}$ onto $\mathcal{Q}/\rho$ such that $h(s) = [s]$ for each $s \in S$.

Any homomorphism h from one monadic algebra $\mathcal{Q}$ onto another $\mathcal{Q}'$ provides us with a rule for defining a congruence relation ρ on $\mathcal{Q}$. We simply take $s\rho s'$ to mean that $h(s) = h(s')$. In this case the quotient algebra $\mathcal{Q}/\rho$ is isomorphic to the image algebra $\mathcal{Q}'$, the isomorphism being given by the rule that each ρ congruence class $[s]$ corresponds to the image $h(s)$.

Equivalence relations on a set, say S, are partially ordered by refinement. If ρ and λ are two equivalence relations, then we say ρ is a <u>refinement</u> of λ, written $\rho \le \lambda$ if whenever $s\rho s'$ then $s\lambda s'$. Since congruence relations are also equivalence relations they are also partially ordered by refinement.

The partially ordered set of equivalence relations on a set S forms what is called a complete lattice. This means that for every nonempty set Z of equivalence relations on S there are equivalence relations $\cap Z$ and $\cup Z$ which are greatest lower and least upper bounds respectively to the set Z. Of particular interest is the fact that if Z consists exclusively of congruence relations, then $\cap Z$ and $\cup Z$ are also congruence relations. Thus, the set of all congruence relations on an algebra α forms a complete sublattice of the complete lattice of equivalence relations on S.

An important result may now be stated. If ξ is an equivalence relation on the set S of states of a monadic algebra α, then there is a unique maximum congruence relation $\tau(\xi)$ which is a refinement of ξ. To prove the above statement, we define K to be the set of all congruence relations ρ which are refinements of ξ. It is clear that K is nonempty since the identity relation s = s' is trivially a refinement of ξ and a congruence relation. Then UK is a congruence relation which is a refinement of ξ. Since every member of K is a refinement of UK we see that it satisfies the requirements of $\tau(\xi)$.

Exercise 16: If ρ is a congruence relation on a monadic algebra α, show that whenever $s\rho s'$ and $t \in T$, then $st\rho s't$.

Exercise 17: Let $\alpha = (\Xi,S,M)$ be a monadic algebra with a distinguished initial state s_0. Show that there is a unique congruence relation ρ on $\mathfrak{F}(\Xi)$ such that $\mathfrak{F}(\Xi)/\rho$ is isomorphic to α and $[\Lambda]$ corresponds to s_0.

6. Define a recognition device by giving a monadic algebra $\alpha = (\Xi,S,M)$ with a distinguished initial state s_0 and specifying a subset F of S called the set of final states. A convenient way of visualizing a recognition device is to imagine an automaton which is capable of receiving a sequence a(0)a(1)...a(r) of elements of Ξ on its input channel. Its output channel, however, is replaced by a simple indicator light. This light is on when the machine is in a state of F and is off otherwise.

D. E. Muller

If $t \in T$, then we say t is <u>accepted</u> by the recognition device $(\mathcal{Q}, F)$

if $s_0 t \in F$. To clarify what acceptance means, let us suppose that t is

the string $a(0)a(1)\ldots a(r)$. Then if $\mathcal{Q}$ is regarded as a machine which

is placed in its initial state s_0 at time 0, and we apply inputs

$a(0)a(1)\ldots a(r)$ at times $0,1,\ldots,r$ respectively the automaton will pass

through states $s(0)s(1)\ldots s(r)s(r+1)$ where $s(0) = s_0$ and $s(i+1) =$

$M(s(i),a(i))$ for $i = 0,\ldots,r$. The state $s(r+1)$ reached at time r+1 is

just the state $s_0 t$, so if the light is on at time r+1 this means that

$s_0 t \in F$ and the string t is accepted.

The set of all strings accepted by $(\mathcal{Q},F)$ is called the set

<u>recognized</u> by $(\mathcal{Q},F)$ and is written $\mathcal{T}(\mathcal{Q},F)$. In theorem 6, we saw that

there is a homomorphism h from $\mathcal{F}(\Xi)$ onto $\mathcal{Q}$ such that $h(t) = s_0 t$ for

each $t \in T$. Thus, t is accepted iff $h(t) \in F$ and $\mathcal{T}(\mathcal{Q},F) = h^{-1}(F)$.

When we let $\mathcal{Q}$ be a groupoid rather than a monadic algebra, a

similar definition may be written. We may again define a recognition

device as a pair $(\mathcal{Q},F)$, where $\mathcal{Q}$ is a groupoid with a distinguished

generator s_0 and F is a subset of the set S of states of $\mathcal{Q}$.

A result analogous to theorem 6 holds in the case of groupoids,

see exercise 8. A unique homomorphism h from the free groupoid $\mathcal{F}_1$

generated from ϕ onto $\mathcal{Q}$ exists which satisfies $h(\phi) = s_0$. Now, we say

that an element t of $\mathcal{F}_1$ is <u>accepted</u> by $(\mathcal{Q},F)$ iff $h(t) \in F$. Also, we

define $\mathcal{T}(\mathcal{Q},F) = h^{-1}(F)$ as the set <u>recognized</u> by $(\mathcal{Q},F)$.

We shall say that a set is <u>regular</u> if it is recognized by a

recognition device with a finite set of states.

<u>Theorem 10</u>: There is unique minimum recognition device $(\mathcal{Q},F)$ which

recognizes any subset T' of T. That is, if $(\mathcal{Q}',F')$ is any other

recognition device which also recognizes T', then $\mathcal{Q}$ must be a homomorphic image of a' under a homomorphism h such that $F' = h^{-1}(F)$.

Proof: Define an equivalence relation ξ on T by letting $t\xi t'$ mean $t \in T \Leftrightarrow t' \in T$. Thus, ξ produces the partition $\{T', T-T'\}$ on T. We have seen that $\tau(\xi)$ is the maximum congruence relation on $\mathfrak{J}(\Xi)$ which is a refinement of ξ. Thus define a as $\mathfrak{J}(\Xi)/\tau(\xi)$ and F as the set of all $\tau(\xi)$ congruence classes included in T'. It is trivial to see that (a, F) recognizes T'.

Now, suppose that (a', F') also recognizes T'. Let h be the unique homomorphism from $\mathfrak{J}(\Xi)$ onto a'. It yields a congruence relation μ by the rule $t\mu t' \Leftrightarrow h(t) = h(t')$. Since μ is a refinement of ξ it is also a refinement of $\tau(\xi)$. Thus, there is a homomorphism from $\mathfrak{J}(\Xi)/\mu$ onto $\mathfrak{J}(\Xi)/\tau(\xi)$. Since a' is isomorphic with $\mathfrak{J}(\Xi)/\mu$ and a with $\mathfrak{J}(\Xi)/\tau(\xi)$ the result follows. $\square$

We see that a subset T' of T is regular if and only if the minimum recognition device which recognizes it has a finite set of states. This device, then, has fewer states than any other recognition device which recognizes T' since it is a homomorphic image of any such device. Since T' is a union of $\tau(\xi)$ classes, we also see that T' is regular if and only if it is the union of some of the congruence classes of a congruence relation having a finite number of congruence classes.

In practice, we may describe a regular set by giving the description of a recognition device which recognizes it. If this device has a finite set of states we may wish to know whether it is the unique device of theorem 10 which has a minimum number of states. If it is not then we may wish to find the device which does.

Since by theorem 10 the device (a, F) which has a minimum number of states is a homomorphic image of any other device (a', F') which

D. E. Muller

recognizes the same set T', we may use this homomorphism to define a
congruence relation ρ on $\mathcal{A}'$ such that $\mathcal{A}'/\rho$ is isomorphic with $\mathcal{A}$. The
states of $\mathcal{A}$ then correspond to the ρ congruence classes of $\mathcal{A}'$. Thus,
one way to obtain $(\mathcal{A},F)$ if we are given $(\mathcal{A}',F')$ is to determine the
congruence ρ, and this is done in the following algorithm.

Algorithm: (Moore) Given a recognition device $(\mathcal{A}',F')$, where $\mathcal{A}'$
is a monadic algebra (Ξ,S',M') with a distinguished initial state s_0'
and a finite state set S', then we obtain the recognition device $(\mathcal{A},F)$
which recognizes the same set as $(\mathcal{A}',F')$ and has a minimum set of states
by the following steps.

1. Represent the equivalence relation we shall call ρ_0 by
    writing the partition $\{F',S'-F'\}$.

2. For $i \geq 0$, represent each equivalence relation ρ_{i+1} by
    forming the corresponding partition so that $s\rho_{i+1}s'$ whenever

    (i) $s\rho_i s'$, and

    (ii) for all $a \in \Xi$, $sa\rho_i s'a$.

Continue forming new relations ρ_i as long as further refinement is
obtained, and stop at step k when ρ_k and ρ_{k+1} are identical. We then
take $\rho = \rho_k$.

It is clear from the definition that ρ_{i+1} is a refinement of ρ_i
for $i = 0,1,\ldots,k-1$. No further applications of step 2 can produce
further refinement after ρ_k. It is also clear since $\rho = \rho_k = \rho_{k+1}$ that
ρ must be a congruence relation by the property (ii). Since S' is
finite ρ_k can have no more classes than $|S'|$, the number of states of
S'. We know that ρ_0 has two equivalence classes and $\rho_k < \rho_{k-1} < \cdots < \rho_0$
so $k \leq |S'|-2$.

Now, let ζ be another congruence relation on S' which is a refinement of ρ_o. Assume inductively that $\zeta \leq \rho_i$. If $s\zeta s'$ then $s\rho_i s'$. Also, since ζ is a congruence relation, if $a \in \Xi$, then $sabs'a$, so $sa\rho_i s'a$. Using (i) and (ii) we see that $s\rho_{i+1}s'$. Hence $\zeta \leq \rho_{i+1}$. By induction we therefore obtain $\zeta \leq \rho_k = \rho$. This shows that ρ is the maximum congruence relation on S' which is a refinement of ρ_o. Since F' must be a union of ρ classes we see that we must start with ρ_o.

The recognition device with a minimum number of states $(\mathcal{C},F)$ which recognizes the same set as $(\mathcal{C}',F')$ is obtained by letting $\mathcal{C}$ be $\mathcal{C}'/\rho$ the monadic algebra of ρ congruence classes and letting F be the family of ρ congruence classes included in F'. We will give one example in which $\mathcal{C}$ is a monadic algebra and a second in which $\mathcal{C}$ is a groupoid.

Example 1: Let $\Xi = \{0,1\}$, $S' = \{s_0', s_1', s_2', s_3'\}$ and let M' be given by the table:

	0	1
s_0'	s_1'	s_2'
s_1'	s_2'	s_3'
s_2'	s_2'	s_3'
s_3'	s_1'	s_2'

, while $F' = \{s_3'\}$.

M'

1. The partition corresponding to ρ_o is

$$\{\{s_0', s_1', s_2'\}, \{s_3'\}\}.$$

2. Forming ρ_1 we obtain the partition

$$\{\{s_0'\}, \{s_1', s_2'\}, \{s_3'\}\},$$

because although $s_0'1 = s_2'$, we have $s_1'1 = s_3'$ and $s_2'1 = s_3'$. The fact that s_2' and s_3' are in different ρ_o classes thus requires that s_0' and s_1' are in different ρ_1 classes.

3. Checking ρ_2 we see that it is the same as ρ_1 and the process may be terminated.

4. To construct $(\mathcal{C}, F)$, we let $s_0 = [s_0']$, $s_1 = [s_1'] = [s_2']$, and $s_2 = [s_3']$. The corresponding table is given by

	0	1
s_0	s_1	s_1
s_1	s_1	s_2
s_2	s_1	s_1

,

and $F = \{s_2\}$.

Example 2: Let $\mathcal{C}'$ be a groupoid whose state set consists of 8 states $\{s_0, \ldots, s_7\}$. The rules obeyed by the binary operation γ are:
$s_0 \gamma s_0 = s_1$, $s_0 \gamma s_1 = s_2$, $s_1 \gamma s_0 = s_3$, $s_3 \gamma s_3 = s_4$, $s_3 \gamma s_4 = s_5$, $s_5 \gamma s_3 = s_6$, $s_3 \gamma s_6 = s_5$, and all other products are s_7. We take $F' = \{s_4, s_6\}$.

1. The partition of ρ_0 is $\{\{s_0, s_1, s_2, s_3, s_5, s_7\}, \{s_4, s_6\}\}$.

2. In carrying out the algorithm for a groupoid, we must replace step 2 of the algorithm with the following.

2 . For $i \geq 0$, represent each equivalence relation ρ_{i+1} by forming the corresponding partition so that $s \rho_{i+1} s'$ whenever

(i) $s \rho_i s'$ and

(ii) for all $s'' \in S'$, $s \gamma s'' \rho_i s' \gamma s''$ and $s'' \gamma s \rho_i s'' \gamma s'$.

The partition of ρ_1 may be seen to be $\{\{s_0, s_1, s_2, s_7\}, \{s_3\}, \{s_4, s_6\}, \{s_5\}\}$. This new partition is obtained because $s_0 \gamma s_3 = s_7$ and $s_3 \gamma s_3 = s_4$ are in different ρ_0 classes, forcing s_0 and s_3 into different ρ_1 classes. Also, $s_0 \gamma s_3$ and $s_5 \gamma s_3 = s_6$ are in different ρ_0 classes so s_0 and s_5 are in different ρ_1 classes. Furthermore, $s_3 \gamma s_3 = s_4$ and $s_3 \gamma s_5 = s_7$ are in different ρ_0 classes, placing s_3 and s_5 in different ρ_1 classes. We can check that all other ρ_1 congruences hold.

3. The partition of ρ_2 is $\{\{s_0,s_2,s_7\},\{s_1\},\{s_3\},\{s_4,s_6\},\{s_5\}\}$.
   We see that $s_1\gamma s_0 = s_3$ and $s_0\gamma s_0 = s_1$ are in different ρ_1
   classes placing s_0 and s_1 in different ρ_2 classes. All other
   ρ_2 may be shown to hold.

4. The partition of ρ_3 in $\{\{s_0\},\{s_1\},\{s_2,s_7\},\{s_3\},\{s_4,s_6\},\{s_5\}\}$.
   In this case, $s_0\gamma s_0 = s_1$ and $s_0\gamma s_2 = s_7$ are in different ρ_2
   classes. This partition is the final one because $\rho_3 = \rho$ is a
   congruence on the original groupoid and no further refinement
   can be made by application of rule 2' of the algorithm. The
   minimum recognition device has 6 states which we may label by
   their earliest representatives. Its operations satisfy the
   following rules, $s_0\gamma s_0 = s_1$, $s_1\gamma s_0 = s_3$, $s_3\gamma s_3 = s_4$,
   $s_3\gamma s_4 = s_5$, $s_5\gamma s_3 = s_4$, and all other products are s_2. The
   set of final states is just the singleton $F = \{s_4\}$.

It is interesting to note that this device accepts members of $\mathfrak{I}_1$ rather than members of $\mathfrak{I}(\Xi)$ as is the case when $\mathcal{Q}$ is a monadic algebra. The members of $\mathfrak{I}_1$ are called <u>binary trees</u>. They can be represented in various ways but two of the most convenient are as formulas involving the generator ϕ and the operation γ and as graphs. For example, we may write the formula $t = ((\phi\gamma\phi)\gamma((\phi\gamma\phi)\gamma\phi))\gamma(\phi\gamma\phi)$ to represent a tree. To determine whether this tree t is accepted by the final recognition device we apply the homomorphism h, which involves substituting s_0 for each appearance of ϕ and using the rules of M. $h(t) =$

$((s_0\gamma s_0)\gamma((s_0\gamma s_0)\gamma s_0))\gamma(s_0\gamma s_0) = (s_1\gamma(s_1\gamma s_0))\gamma s_1 = (s_1\gamma s_3)\gamma s_1 =$

D. E. Muller

$s_2 \gamma s_1 = s_2$. Since $s_2 \notin F$ we see that this tree is not accepted. On the other hand, we may check that the tree $((\phi\gamma\phi)\gamma\phi)\gamma((\phi\gamma\phi)\gamma\phi)$ is accepted.

To represent a tree as a graph, we let ϕ be a single node. When t_1 and t_2 are represented as trees, then inductively $t_1 \gamma t_2$ is drawn as shown in figure 2a below. The graph representing the tree t above is shown in figure 2b. In figure 2b, each node of the graph is labeled with the image under h of the subtree of which it is the root. This label is always a state of S, and to determine whether or not the tree is accepted we simply check to see whether the label of the root, or topmost node is F.

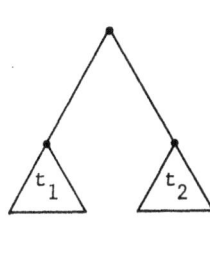

a

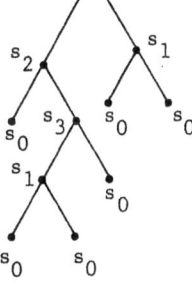

b

Figure 2

Aleshin, S. V., "Finite automata and Burnside's problem for periodic groups," Matematicheskie Zametki, 11, 3, pp. 319-328, (1972).

Eilenberg, Samuel, Automata, Languages and Machines, Academic Press, (1974).

Ginsburg, Seymour, An Introduction to Mathematical Machine Theory, Addison-Wesley, (1962).

Ginzburg, Abraham, Algebraic Theory of Automata, Academic Press, (1968).

Hartmanis, J. and Stearns, R. E., Algebraic Structure Theory of Sequential Machines, Prentice-Hall, (1966).

Kohavi, Zvi, Switching and Finite Automata Theory, McGraw-Hill, (1970).

Miller, Raymond E., Switching Theory, in two volumes, John Wiley and Sons, (1965).

Rabin, Michael O., Automata on Infinite Objects and Church's Problem, American Math. Society, (1972).

Salomaa, Arto, Theory of Automata, Pergamon Press, (1969).

Trakhtenbrot, B. A., and Barzdin, Ya.M., Finite Automata, Behavior and Synthesis, North-Holland, (1973).

Turing, A. M., "On computable numbers, with an application to the Entscheidigungsproblem," Proc. London Math. Soc. 42, 230-265, (1936).

Von Neumann, J., "Probabilistic logics and the synthesis of reliable organisms from unreliable components," Automata Studies, pp. 43-98, Princeton University Press, Princeton, N.J., (1956).

CENTRO INTERNAZIONALE MATEMATICO ESTIVO
(C.I.M.E.)

COMPUTATIONAL COMPLEXITY OF COMBINATORIAL
AND GRAPH-THEORETIC PROBLEMS

R.M. KARP

Corso tenuto a Bressanone dal 9 al 17 giugno 1975

R. M. Karp

COMPUTATIONAL COMPLEXITY OF COMBINATORIAL

AND GRAPH-THEORETIC PROBLEMS

R.M. Karp

University of California, Berkeley

1. Basic Terminology

This series of lectures is concerned with efficient algorithms which
operate on graphs. A graph is a structure consisting of a finite set of
vertices, certain pairs of which are joined by edges. In an ordinary
graph an edge has no direction specified. In a digraph each edge is
directed from one vertex to another. Figure 1 indicates a graph and a
digraph.

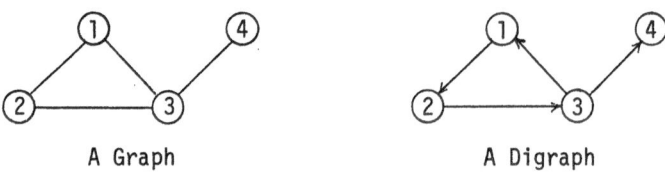

A Graph   A Digraph

Figure 1

The underlying structure of many different types of objects can be
represented by graphs or digraphs. Thus a computer program can be represented by a digraph in which the vertices represent straight-line
sequences of code and the edges represent the branching structure of the
program. Data structures can be represented by digraphs with a vertex
for each field of data and an edge for each link between fields. Graphs
or digraphs can also represent finite-state systems, transportation networks, electric networks, chemical molecules, social structures, etc.

We begin with some elementary definitions concerning graphs. A graph

R. M. Karp

$G = (V,E)$ is specified by a finite set V of vertices and a finite collection E of two-element sets of vertices called edges. A graph
$G' = (V',E')$ is called a subgraph of G if $V' \subseteq V$ and $E' \subseteq E$. If
$\{x,y\}$ is an edge then vertices x and y are said to be adjacent, and
to be incident with the edge $\{x,y\}$. A sequence of distinct edges of the
form $\{x_1,x_2\},\{x_2,x_3\},\{x_3,x_4\},\ldots,\{x_{n-1},x_n\}$ is called a path. The path
is simple if $x_1,x_2,\ldots,x_{n-1}$ are all distinct, and a cycle if $x_1 = x_n$.
The null sequence of edges is by convention both a path and a cycle.

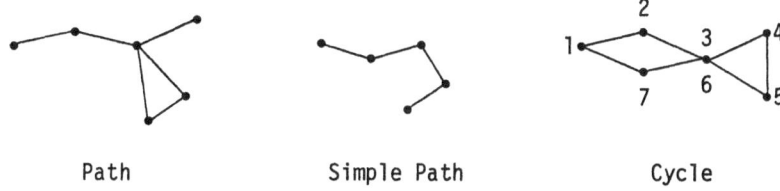

| Path | Simple Path | Cycle |

A graph G is connected if every pair of vertices is joined by a
path. A component of G is a maximal (with respect to inclusion) connected subgraph of G. The following graph has four components.

A tree is a connected graph without cycles; for example,

A spanning tree of $G = (V,E)$ is a subgraph $T = (V,E')$ which is a
tree and includes all the vertices of G.

Exercise. G is connected if and only if G has a spanning tree.

The Representation of Graphs

To manipulate graphs algorithmically we need a precise way of representing them as strings of symbols; the pictures we have been using are
suitable for humans, but not for computers. Normally we represent a
graph by giving a singly linked list of its vertices and, for each vertex
x, a singly linked list of the vertices adjacent to x. We call this
the adjacency list representation. The adjacency matrix is another way
to represent a graph. It is a $|V| \times |V|$ matrix of 0's and 1's; the x,y
entry in this matrix is 1 if and only if {x,y} is an edge. It follows
that the adjacency matrix of a graph is symmetric. Similar representations can be defined for digraphs. The adjacency matrix of a digraph is
in general not symmetric.

In practice we are often concerned with "sparse" graphs, in which
there are many vertices, but only a small number of edges incident with
each vertex. The adjacency list representation is more compact than the
adjacency matrix for such graphs. Because of this and other considerations of efficiency we normally prefer the adjacency list representation.

Searching a Graph

We outline an algorithm for finding the vertex set C of the component of G containing a given vertex v. The algorithm also produces the
edge set T of a spanning tree of this component.

SEARCH ALGORITHM
C ← {v}
T ← φ
while there is an unexplored edge {x,y} such that x ε C, do
    if y ε C, do nothing

else
    $C \leftarrow C \cup \{y\}$
    $T \leftarrow T \cup \{x,y\}$

Note that an edge is added to the spanning tree T precisely when a new

vertex is added to the component C.

2.   Depth-First Search and Its Applications [58]

The search algorithm given above is not fully specific, since it does

not tell us which unexplored edge to choose. We discuss here a particu-

larly convenient and useful form of the algorithm called depth-first

search, in which, at any given time, the last vertex added to the compo-

nent C is the first to have its adjacency list explored. The implemen-

tation of depth-first search uses a data structure called a pushdown

stack. A pushdown stack is a list which we always manipulate at its

right-hand end. We can PUSH a new element onto the right-hand end or POP

the rightmost element from the stack. The rightmost element is called the

TOP of the stack. At any stage in a depth-first search, the stack con-

tains all those vertices which have been reached, but whose adjacency list

has not been fully explored. The depth-first search also creates a set T

of directed edges called tree edges and a set F of directed edges called

fronds.

Depth-First Search

    PUSH v; $C \leftarrow \{v\}$; STACK $\leftarrow \{v\}$
    while STACK $\neq \phi$ do begin
      if adjacency list (TOP) contains an unexplored edge to vertex y
      then begin
          if $y \notin C$ then $T \leftarrow T \cup \{(TOP,y)\}$; $C \leftarrow C \cup \{y\}$;
                  STACK $\leftarrow$ STACK $\cup \{y\}$; PUSH y
            else if $y \in$ STACK then $F \leftarrow F \cup \{(TOP,y)\}$

R. M. Karp

      <u>end;</u>
    <u>else</u> STACK ← STACK - {TOP}; POP; <u>end;</u>

The depth-first search algorithm PUSHes each reachable vertex onto the stack once, and explores each edge once in each direction. Thus its running time is $O(|V| + |E|)$. The edges in T form a spanning tree of the component containing v, with each edge directed away from the root; the fronds are directed toward the root, i.e., from a descendant to an ancestor.

To illustrate the method, we perform a depth-first search of the following graph. The vertices are numbered in the order of their first occurrence on the pushdown stack.

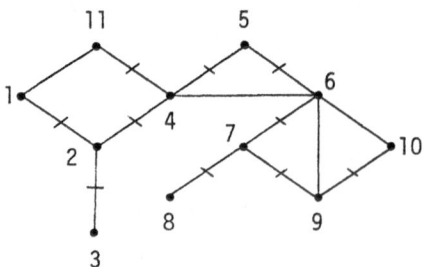

The tree edges are marked. Orienting the tree edges and fronds as indicated in the algorithm, we get:

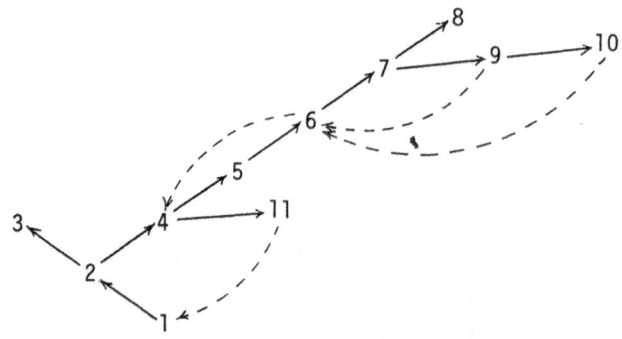

R. M. Karp

Every frond runs from a vertex in the tree to one of its ancestors. Notice that depth-first search has the following special property: if {x,y} is an edge of the graph, then, in the tree resulting from depth-first search, either y is an ancestor of x or x is an ancestor of y.

Let G be a connected graph. Define a cut vertex of G as a vertex whose removal, together with the removal of all edges incident with it, results in a disconnected graph. In the foregoing example, the cut vertices are 4, 6 and 7. Define a block as a maximal subgraph having no cut vertices. The blocks are:

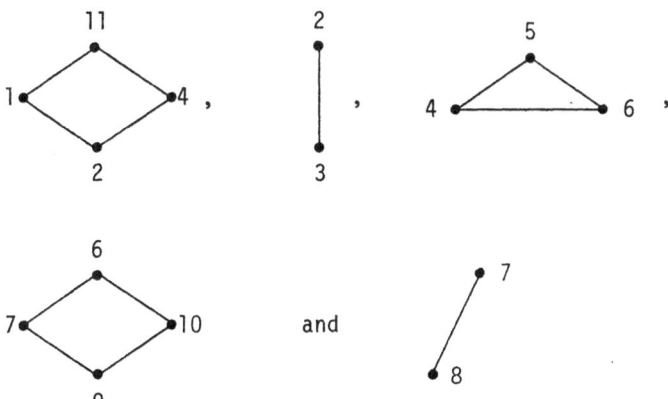

We show how the structure created by depth-first search can be used to determine the cut vertices and blocks of a graph. The structure is a digraph with two types of edges, called tree edges and fronds. The vertices are numbered in the order in which they have been pushed onto the stack. Note that the sequence of numbers on any path directed away from the root is increasing. The following figure will be used for illustration.

R. M. Karp

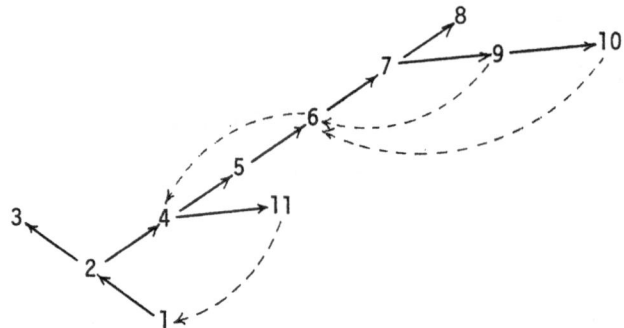

Throughout this discussion we will not distinguish between a vertex and the number assigned to it. For any vertex x, let Tree(x) denote the vertices reachable from x by a single tree edge, and let Frond(x) denote the vertices reachable from x by a single frond. In the example Tree(7) = {8,9}, Tree(8) = φ and Frond(9) = {6}. Also, define Anc(x) as those vertices (including x) which are in the unique path in the tree from the root to x. Thus, Anc(5) = {1,2,4,5}.

We define a function which will play a key role in the test for cut vertices. Define Low(x) as the least vertex which is either x itself or is reachable from x by a (possibly null) sequence of directed tree edges followed by one frond. For example, Low(5) = 4, Low(11) = 1, Low(7) = 6. Note that Low(x) ε Anc(x).

We see that

$$Low(x) = min[\{x\} \cup Frond(x) \cup Low(Tree(x))] .$$

This expression permits the recursive evaluation of the function Low. This evaluation can be performed in one pass through the vertices, but this pass must always inspect the vertices in Tree(x) before it inspects x. Such an ordering is POP order; i.e., the order in which the vertices are POPed from the stack during the original depth-first search.

Since POP order has the desired property, the computation of Low can be integrated with the original depth-first search. Low(x) is evaluated precisely when x is POPed from the stack. In the example, the function Low(x) is:

x	1	2	3	4	5	6	7	8	9	10	11
Low(x)	1	1	3	1	4	4	6	8	6	6	11

Next we show that the function Low(x) permits easy identification of the cut vertices.

Theorem. Let x be a vertex which is not the root of the tree. Then x is a cut vertex if and only if, for some $y \in$ Tree(x), Low(y) $\geq$ x.

Proof. Let Tree(x) = $\{y_1, y_2, \ldots, y_p\}$. Removal of x from the tree leaves p+1 components; the subtrees rooted at $y_1, y_2, \ldots, y_p$, and the "main component", containing those vertices which are not descendants of x.

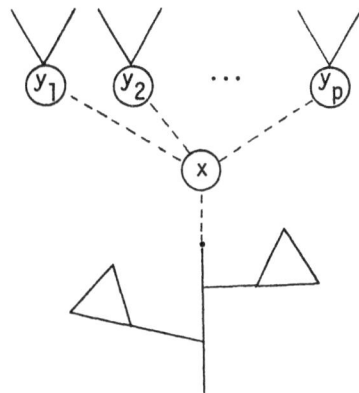

If each of the p subtrees has a way to reach the main component without passing through x, then clearly removal of x leaves a connected graph.

This is, in fact, a necessary and sufficient condition. But the subtree
rooted at y_i can reach the main component only by a frond from some
descendant of y_i to the main component; and such a frond exists precisely when $Low(y_i) < x$. ☐

Having tabulated $Low(\cdot)$, we may take each vertex y in turn and
test whether $Low(y) \geq x$, where $y \in Tree(x)$. The cut vertices are thus
determined:

y	$Low(y)$	x	
3	3	2	
5	4	4	cut
7	6	6	vertices
8	8	7	

The entire computation requires one pass through the structure, and hence
$O(|V| + |E|)$ steps.

Digraphs

It is possible to carry out a depth-first search of a digraph, and
thereby visit all the vertices reachable from some starting vertex. The
process is complicated by the fact that the original order of the edges
has to be respected. Because of this, two types of edges occur besides
the tree edges (directed from father to son) and the fronds (directed
from descendant to ancestor). These are the reverse fronds (directed
from an ancestor to a descendant) and the cross-links. A cross-link
(x,y) joins two vertices which are incomparable in the tree; i.e.,
neither is an ancestor of the other. Whenever such a cross-link occurs,
we have $x > y$, where the vertices are numbered in PUSH order. The fact
that cross-links always run from a higher numbered vertex to a lower-

R. M. Karp

numbered one is a special property of depth-first search. Below we show
a digraph, and exhibit the result of a depth-first search.

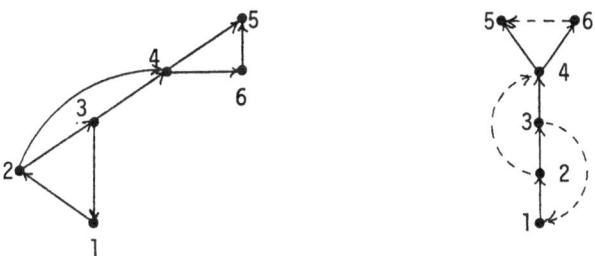

The edge (2,4) is a reverse frond, and (6,5) is a cross-link.

   We shall use depth-first search as the basis of an algorithm to find
the strong components of a digraph. A digraph is strongly connected if,
given any two distinct vertices x and y, there is a path from x to
y. Thus, if we think of the vertices of a digraph as representing the
states of a system, and the edges as representing transitions, then strong
connectedness means that, no matter what state the system is in, every
state remains reachable. A strong component of a digraph is a maximal
strongly connected subgraph. The strong components of the following
digraph are circled.

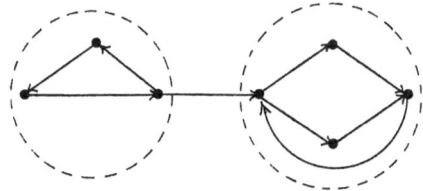

Note that every vertex is in exactly one strong component, and that an
edge is in a strong component if and only if it is part of a directed
cycle.

   Let G be a digraph. For any vertex x, let R(x) denote the set

R. M. Karp

of vertices reachable from x. In particular, the null path takes x to itself, so $x \in R(x)$. We define the equivalence relation $\cong$ by: $x \cong y$ if and only if x and y are in the same strong component of G. It is an easy exercise to show that $x \cong y$ if and only if $R(x) = R(y)$.

We now discuss the computation of the strong components of G. For convenience assume there is a vertex s from which every vertex is reachable (we leave it as an exercise to extend the discussion to the case where such an s does not exist). Starting at s, conduct a depth-first search of G. Recall that $Anc(x)$ denotes the set of ancestors of x, including x itself.

Lemma. $x \cong y \Leftrightarrow R(x) \cap R(y) \cap Anc(x) \cap Anc(y) \neq \phi$.

Proof. ($\Rightarrow$) We consider two cases. Either x and y are comparable in the tree or they are incomparable. In the former case, suppose without loss of generality that $x \in Anc(y)$. Then $x \in R(x) \cap R(y) \cap Anc(x) \cap Anc(y)$. Now suppose they are incomparable, and y has a higher number than x. Then the tree looks as follows, where z is the highest common ancestor of x and y. Since all cross-links run from right to left, any path

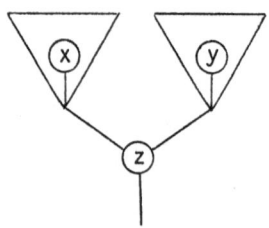

from the subtree containing x to the one containing y must pass through an element of $Anc(z)$. Hence $z \in R(x) \cap R(y) \cap Anc(x) \cap Anc(y)$.

($\Leftarrow$) Let $t \in R(x) \cap R(y) \cap Anc(x) \cap Anc(y)$. Then $\{x,y\} \subseteq R(t)$, so $x \in R(y)$ and $y \in R(x)$; i.e., there is a path from x to y through t,

R. M. Karp

and a path from y to x through t. Hence $x \cong y$. □

Now define Bottom(x) as the lowest numbered ancestor of x reachable from x. Bottom(x) = min[R(x) ∩ Anc(x)].

Theorem. $x \cong y$ ⟺ Bottom(x) = Bottom(y) .

Proof. (⟹) Since $x \cong y$, R(x) = R(y), and R(x) ∩ Anc(x) ∩ Anc(y) ≠ φ.
Then Bottom(x) = min(R(x) ∩ Anc(x)) = min(R(x) ∩ Anc(x) ∩ Anc(y)) =
= min(R(x) ∩ R(y) ∩ Anc(x) ∩ Anc(y)). But, by the same argument applied to
y, this quantity is also equal to Bottom(y).

(⟸) If Bottom(x) = Bottom(y) then Bottom(x) ε (R(x) ∩ Anc(x)) ∩
(R(y) ∩ Anc(y)). Thus, by the lemma, $x \cong y$. □

In view of this theorem, we can determine the strong components by
tabulating the function Bottom(·). We do so by inspection in the following example.

Example.

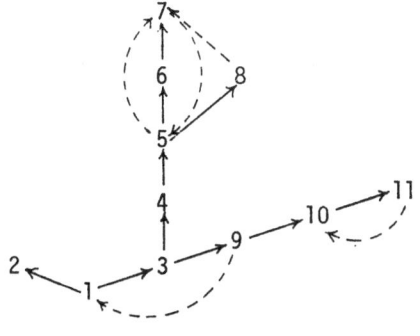

x	1	2	3	4	5	6	7	8	9	10	11
Bottom(x)	1	2	1	4	5	5	5	5	1	10	10

We show that it is possible to tabulate the function Bottom(·) by a
process that makes three passes through the vertices and edges of the

graph. This yields an algorithm whose running time is $O(|V| + |E|)$. We have already defined the functions Tree(x), Frond(x) and Anc(x). In the same spirit, define

$$Cross(x) = \{v|\ (x,v)\ \text{is a cross-link}\}\ .$$

We will also need the following auxiliary definitions. Low(x) is the minimum of x, and of all vertices reachable by a path consisting of tree edges followed by exactly one frond. $Low^*(x)$ is the least vertex reachable from x by a path consisting of any number of tree edges and fronds. Side(x) is the least ancestor of x reachable by a path in which a cross-link occurs before any frond. Then

$$Low(x) = \min[\{x\} \cup Frond(x) \cup Low(tree(x))]$$
$$Low^*(x) = \underline{if}\ Low(x) = x\ \underline{then}\ x\ \underline{else}\ Low^*(Low(x))$$
$$Side(x) = \min[\{x\} \cup Side(Tree(x))$$
$$\cup\ (Bottom(Cross(x)) \cap Anc(x))]$$
$$Bottom(x) = Side(Low^*(x))\ .$$

To determine Low(x) we go through the vertices in POP order. Then we compute $Low^*(x)$ by going through in PUSH order, and then simultaneously tabulate Side(x) and Bottom(x) by a final pass in POP order.

3.   Computing Shortest Paths [13],[15]

We next give a well-known algorithm due to Dijkstra which computes the shortest path from a given source to each of the vertices in a digraph whose arcs are assigned nonnegative weights. Let $G = (V,E)$ be a digraph; for convenience, take $V = \{1,2,...,n\}$. Associate with each directed edge (i,j) a weight d_{ij}, which may be interpreted as the

R. M. Karp

cost of traversing the edge. By convention, take $d_{ij} = \infty$ if $(i,j) \notin E$
Define $\pi^*(i)$ as the minimum cost of a path from 1 to i; in particular, $\pi^*(1) = 0$. The algorithm proceeds through n-1 "major cycles".
At the beginning of each such cycle, we will be given a partition of the
vertices into two sets, S and $\bar{S}$, with $1 \in S$. Also, the following
data will have been tabulated:

   for each $i \in S$, $\pi^*(i)$

   for each $k \in \bar{S}$, $\pi(k) = \min_{i \in S}[\pi(i) + d_{ik}]$.

The value $\pi(k)$ gives the minimum cost of a path from 1 to k, subject to the condition that every vertex in the path except k is in S.

   The correctness of the algorithm depends on the following lemma.

Lemma. Let $j \in \bar{S}$ be such that $\pi(j) = \min_{k \in \bar{S}} \pi(k)$. Then $\pi^*(j) = \pi(j)$.

Proof. Clearly there is a path from 1 to j of cost $\pi(j)$. To show
that there is no cheaper path from 1 to j, consider an arbitrary path
P from 1 to j and divide it into two parts: P_1, the part up to and
including the first occurrence in P of a vertex $\ell \in \bar{S}$, and P_2, the
rest of the path. Then the cost of $P_1 \geq \pi(\ell) \geq \pi(j)$, and the cost of
$P_2 \geq 0$. Hence the overall cost of $P \geq \pi(j)$. □

Thus the major cycle of the algorithm has the following simple form:

   Select $j \in \bar{S}$ so that $\pi(j) = \min_{k \in \bar{S}} \pi(k)$.

   $\pi^*(j) \leftarrow \pi(j)$

   $S \leftarrow S \cup \{j\}$

   $\bar{S} \leftarrow \bar{S} - \{j\}$

   For $k \in \bar{S}$, $\pi(k) = \min[\pi(k), \pi(j) + d_{jk}]$.

The last step adjusts each value of $\pi(k)$ to account for the fact that

R. M. Karp

vertex j is now available in S.

The algorithm starts with $S = \{1\}$, $\Pi^*(1) = 0$, $\Pi(k) = d_{1k}$, $k \neq 1$. It terminates when all the vertices have been added to S.

Next we discuss the efficient implementation of Dijkstra's shortest-path algorithm. There are two basic processes to be carried out in each major cycle:

(A) Selection of j such that $\Pi(j) = \min_{k \in \bar{S}} \Pi(k)$.

(B) Computation of $\Pi(k) = \min[\Pi(k), \Pi(j)+d_{jk}]$.

The values $\Pi(k)$ will be stored in a one-dimensional array, and the values d_{jk} in a two-dimensional array. Step (B) requires time proportional to the number of edges incident with j, and thus, summed over the entire computation, requires $O(|E|)$ steps. The time for (A) depends on the data structures we use. Suppose first that the names of the elements of $\bar{S}$ are linked together in a circular list. For example, if $n = 10$ and $\bar{S} = \{3,6,8,9\}$ we might have the following structure:

$\Pi(1)$	$\Pi(2)$	$\Pi(3)$	$\Pi(4)$	$\Pi(5)$	$\Pi(6)$	$\Pi(7)$	$\Pi(8)$	$\Pi(9)$	$\Pi(10)$
--	--	6	--	--	8	--	9	3	--

The effect of the indicated link fields is to create the structure

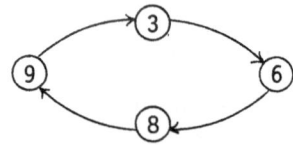

With this representation, a program can cycle through the circular list of elements of $\bar{S}$, find the minimum of the corresponding Π-values, and then eliminate the appropriate element. For example, if $\Pi(6) < \min\{\Pi(3),\Pi(8),\Pi(9)\}$, the structure becomes

R. M. Karp

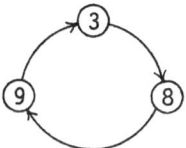

The time to execute step (A) is proportional to $|\bar{S}|$, and hence, summed over the entire process, is

$$O\big((n-1) + (n-2) + (n-3) + \cdots + 1\big) = O(n^2) \ .$$

With this implementation, the running time of the algorithm is $O(|E|) + O(n^2)$. In dense graphs, where $|E| \sim n^2$, the second term is tolerable; but in sparse graphs the second term tends to dominate, and we can improve upon it by using a different data structure. A priority queue is a structure used to represent a set of items, each of which consists of a name and a numerical value. The operations which can be performed on the structure are:

insert a new item

delete the item with a given name

delete the entry with least value

It is possible to implement a priority queue so that each operation takes $O(\log n)$ primitive steps (instruction executions on a random-access machine, for example), where n is the number of items in the structure.

In our application the values $\Pi(k)$, $k \in \bar{S}$, are stored in an array; in addition, there is a priority queue which stores the pairs $(k, \Pi(k))$, $k \in \bar{S}$ [Of course, we avoid redundant storage of $\Pi(k)$ in two places by actually keeping only the values k in the priority queue]. Then the major cycle is as follows:

delete the pair $(j, \Pi(j))$ such that $\Pi(j)$ is least;
mark j to indicate that $j \in S$;

for each k such that $(j,k) \in E$, and $k \notin S$
   compute $\Pi(k) = \min\left(\Pi(k), \Pi(k)+d_{jk}\right)$
if this changes the value of $\Pi(k)$, delete the old item named k,
   and insert a new item $(k,\Pi(k))$.

With this implementation the work for step (B) is $O(\log n)$ steps per edge, and the running time of the algorithm is $O(|E|\log n)$; this is an improvement on the preceding implementation unless the given digraph is very dense.

4.   Computing Maximum Flows [14],[19]

We next consider an important problem in which the edges of a network are regarded as channels along which some commodity can flow (gas, water, electricity, automobiles, bits of information, etc.), and we wish to maximize the rate at which the commodity flows from a given source to a given destination. The data for the problem are:

a digraph $G = (V,E)$

a source $s \in V$

a destination $t \in V$

for each edge (u,v), a capacity $c(u,v) > 0$

The capacity $c(u,v)$ is an upper bound on the rate at which the commodity can flow along the arc (u,v).

We define a flow as any assignment of values to the edges such that

$$0 \leq f(u,v) \leq c(u,v) \qquad (u,v) \in E$$

and

$$f(\{u\},V) = f(V,\{u\}) \qquad u \in V - \{s,t\}$$
$$\text{(Conservation Law)}$$

Here we have used the following notation: for any $X \subseteq V$ and $Y \subseteq V$,

R. M. Karp

$$f(X,Y) = \sum_{\substack{u \varepsilon X \\ v \varepsilon Y}} f(u,v) \ .$$

Thus, we require that the total flow leaving any vertex except s or t
is equal to the total flow entering that vertex.

The following is an easy consequence of conservation. Let $(X,\bar{X})$
be any partition of V into two sets such that $s \varepsilon X$ and $t \varepsilon \bar{X}$. Then

$$f(X,\bar{X}) - f(\bar{X},X) = f(s,V) - f(V,s) = f(V,t) - f(t,V) \ .$$

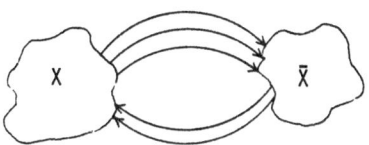

This common value we call Φ, the _value_ of the flow. It represents the
rate at which the network is carrying the commodity from s to t. Note
that

$$\Phi = f(X,\bar{X}) - f(\bar{X},X) \le c(X,\bar{X}) - 0 = c(X,\bar{X}) \ .$$

Thus, the value of the flow is bounded above by the total capacity on all
the arcs from X to $\bar{X}$. Such a set of arcs is called a _s-t cut_. The
max-flow min-cut theorem states that the maximum value of a flow is equal
to the minimum capacity of a s-t cut:

$$\max_{\text{all flows}} \Phi = \min_{\substack{\text{all cuts} \\ (X,\bar{X})}} c(X,\bar{X}) \ .$$

Now we shall prove the max-flow min-cut theorem. We shall do so by
giving a procedure for computing a maximum flow. As a by-product of the
procedure, we shall show that the value of this flow is equal to the

R. M. Karp

capacity of an s-t cut. This will establish that

$$\max_{\text{flows } f} \Phi \geq \min_{\substack{\text{all s-t cuts} \\ X, \bar{X}}} c(X, \bar{X}) .$$

But we have previously established the reverse inequality, and thus it will follow that

$$\max_{\text{flows } f} \Phi = \min_{\substack{\text{all s-t cuts} \\ X, \bar{X}}} c(X, \bar{X}) .$$

First we show that computing a maximum flow is not quite as simple as one might first assume. In the following network, suppose we send 5 units of flow along the edge (s,A), and then divide it so that four units flow along the path A,B,t, and one unit flows along the path A,D,t. Then we get the flow pattern indicated by the circled numbers.

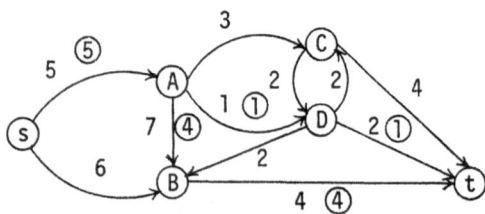

The edges s,A and B,t are saturated, so that there is no way to send more flow along any directed path from s to t. And yet this is not a maximum flow, since the next figure indicates a flow of value 8.

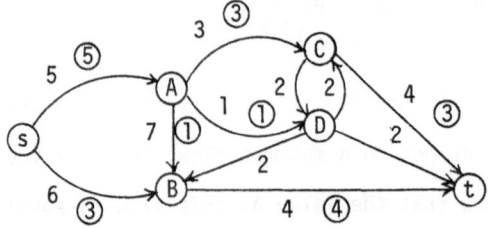

R. M. Karp

This second flow is maximum, since its value is 8, which is equal to the capacity of the s,t cut $(X,\bar{X})$, where

$$X = \{s,A,B\} \quad \text{and} \quad \bar{X} = \{C,D,t\} .$$

Comparing the second flow with the first, we see that it was necessary to decrease (or cancel) three units of flow on the edge (A,B) in order to get a flow of greater value. Thus, any way to compute a maximum flow by successive flow augmentations must permit the possibility of cancelling some flow.

Consider an edge (u,v) with capacity c and flow f. Suppose $0 < f < c$. Then it is possible to increase the flow by as much as $c - f$, or to decrease it by as much as f. We can represent these possibilities by a pair of edges in a new digraph, which we call the augmentation network. There will be a forward edge from u to v labelled $c - f$, and a backward edge from v to u, labelled f.

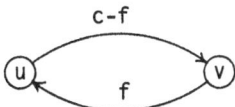

Similarly, if $f = c$, we would have no forward edge, since f cannot increase, but there would be a backward edge labelled c. And finally, if $f = 0$, no backward edge will occur.

The augmentation network corresponding to the first flow above is:

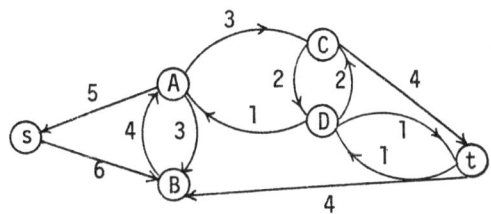

The directed s,t paths in the augmentation network represent possible flow changes that will increase the flow value. The smallest number in any such path represents the amount of flow change possible along the path. Thus the path s, B, A, C, t supports a flow change of 3, giving the second flow shown above. Since the edge B, A in that path is a reverse edge, the flow change cancels three units of flow on the edge A, B, reducing that flow to 1.

Exercise. Prove that a flow change along an s-t path in the augmentation network produces a flow of larger value.

Next we show that, if no s-t path exists in the augmentation network, then the present flow is maximum. Let X denote s, together with all vertices reachable from s by paths in the augmentation network. Then $t \in \bar{X}$, and no edge of the augmentation network crosses from X to $\bar{X}$. Thus, in the original network, every edge (u,v) from X to $\bar{X}$ must be saturated (i.e., f(u,v) = c(u,v)), and every edge from $\bar{X}$ to X must carry no flow; for otherwise an edge from X to $\bar{X}$ would occur in the augmentation network. This means that $f(X,\bar{X}) = c(X,\bar{X})$ and $f(\bar{X},X) = 0$, so

$$\Phi = f(X,\bar{X}) - f(\bar{X},X) = c(X,\bar{X}) \ .$$

Thus Φ must be a maximum flow.

We are now ready to prove the max-flow min-cut theorem. Notice that the process of computing successive augmentations uses only the operations of addition and subtraction. Thus every number computed is an integer if the initial capacities are integer and we start with the zero flow. Hence each augmentation increases Φ by at least one at each iteration, and the process must terminate with a flow whose value is equal to the

R. M. Karp

capacity of some s-t cut. A similar argument applies when the capacities are rational. Thus, when the data is rational, the max-flow min-cut theorem holds. But max Φ and $\min\limits_{X,\bar{X}} c(X,\bar{X})$ are both continuous functions of the data, and hence equality must hold for real data as well. This completes the proof of the max-flow min-cut theorem.

The following example shows that the wrong choice of flow augmenting paths can cause the algorithm to perform badly.

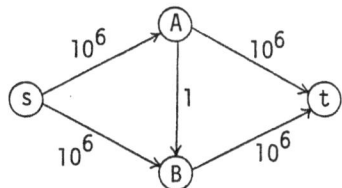

The value of a maximum flow is obviously 2×10^6, but the computation can take 2×10^6 iterations if we insist on augmenting alternately along the paths s,A,B,t and s,B,A,t. The choice of these paths seems unnatural, of course, since they are longer and permit smaller augmentations than the paths s,A,t and s,B,t. Next we will investigate suitable rules for the choice of flow augmenting paths.

We consider a method of computing maximum flows in which, at each step, a shortest augmenting path (i.e., one with fewest edges) is chosen. We show that the total running time of a max-flow algorithm based on this choice is $O(|E| \cdot |V|^2)$.

Suppose we start with some initial flow f^0, and by repeated augmentations along shortest augmenting paths, determine a sequence of flows $\{f^k\}$, each of larger value than its predecessor. Let the augmentation network used in passing from f^k to f^{k+1} be denoted $G^k = (V, E^k)$, and let p^k denote the path from s to t in G^k along which augmentation takes place. For any vertex u, let $\pi^{\wedge}(u)$ denote the length of a shortest path from s to u in G^k, and let $\sigma^k(u)$ denote the length of the shortest path from u to t in G^k. Then

$$\text{length of } P^k = \pi^k(t) = \sigma^k(s) .$$

Also, if (w,x) is an edge in P^k, then $\pi^k(x) = 1 + \pi^k(w)$.

<u>Lemma.</u> For all u, $\pi^{k+1}(u) \geq \pi^k(u)$.

<u>Proof.</u> Consider any edge (x,w) in a shortest path from s to u in G^{k+1}. There are two cases:

(a) $(x,w) \varepsilon E^k$, so $\pi^k(w) \leq 1 + \pi^k(x)$

(b) $(w,x) \varepsilon P^k$, so $\pi^k(x) = 1 + \pi^k(w)$, and $\pi^k(w) < 1 + \pi^k(x)$.

Summing these inequalities along the s-u path, we get

$$\pi^k(u) = \pi^k(u) - \pi^k(s) \leq \sum_{\substack{(x,w) \text{ in} \\ \text{the path}}} 1 = \pi^{k+1}(u) . \qquad \square$$

By a similar reasoning, one can show that, if an edge (x,w) is used in P^k, and later its reversal (w,x) is used in P^{ℓ}, where $\ell > k$, then

$$\text{length of } P^{\ell} > \text{length of } P^k .$$

Also, $\sigma^{k+1}(u) \geq \sigma^k(u)$. These results are left as exercises.

From these theorems we see that the computation of a maximum flow using shortest augmenting paths breaks into at most $|V| - 1$ stages, during each of which all the augmenting paths found are of equal length. During any given stage, no edge is used in both directions.

Consider the implementation of a single stage, commencing when, for the first time, some augmentation network $G^k = (V, E^k)$ does not allow an augmenting path of length $L-1$. Then we begin a search for augmenting paths of length L. By a breadth-first search (really a special case of Dijkstra's shortest-path algorithm) we successively identify vertices at distance $1,2,\ldots$ from s in G^k, and construct a search digraph $G_L^k = (\bar{V},\bar{E})$ whose vertex set $\bar{V}$ is $\{s,t\}\cup\{u|\ \pi^k(u)<L\}$, and whose edge set $\bar{E} = \{(u,v)|\ u\in\bar{V},\ v\in\bar{V}$ and $\pi^k(v) = \pi^k(u) + 1\}$. This results in a digraph which has no cycles, and breaks naturally into $L+1$ levels, as the following figure (with $L = 4$) suggests.

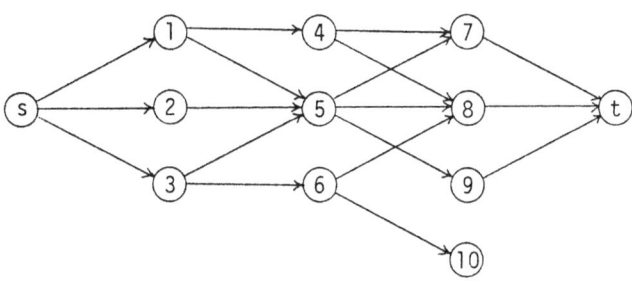

The computation starts by following a sequence of edges starting at s until either t or a dead end is reached. Suppose for instance that the first path found is $s,1,4,7,t$. The successive edges are pushed onto a pushdown stack as they occur. In this case, an augmenting path has been found, since an edge into t was placed on the stack. A flow augmentation is made, and the stack is popped until all edges which saturated at that augmentation have been removed. The saturated edges are no longer useful during the present regime, and are dropped from $\bar{E}$. For example, if edge $(4,7)$ were the only one to saturate, then the stack would be left with $(s,1)(1,4)$, and further search would produce the path $s,1,4,8,t$. If the search terminates with a dead end, then the last edge is popped from the stack and deleted from $\bar{E}$. Altogether $|\bar{E}|$ edges get deleted during one regime, and the number of POP operations or PUSH

R. M. Karp

operations is $O(L|\bar{E}|)$. Thus the number of primitive steps in the entire stage is $O(L|\bar{E}|)$, and summing over all stages, the time is $O(|V|^2|E|)$.

Exercise. Suppose we decide to choose, at each augmentation, the augmenting path that will allow the largest flow change. Discuss how to find such paths, and derive an upper bound on the number of augmentations needed.

5. Matching Theory [3],[16],[18],[22],[26],[32]

A matching in a graph $G = (V,E)$ is a set $M \subseteq E$ such that no two elements of M meet at a common vertex. We consider the problem of finding such a set M with a maximum number of elements.

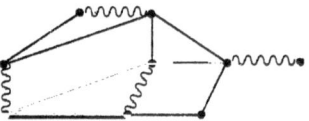

A graph, with the elements of M indicated ($\sim\!\sim$)

An important special case of the matching problem arises when G is bipartite. A graph is bipartite if its vertices can be partitioned into two sets, R and B, such that every edge joins a vertex in R with a vertex in B.

Exercise. G is bipartite if and only if every cycle in G is of even length.

A bipartite graph arises if we take as vertices a set of boys and girls, with a boy and girl joined by an edge if they form a compatible pair. A maximum matching corresponds to a maximum number of disjoint compatible couples.

R. M. Karp

In the bipartite case, the maximum matching problem can be represented as a max-flow problem. Define a digraph $\hat{G} = (\hat{V}, \hat{E})$, where
$\hat{V} = \{s,t\} \cup V$ and $\bar{E} = \{(s,r) \mid r \in R\} \cup \{(b,t) \mid b \in B\} \cup \{(u,v) \mid u \in R, v \in B$
and $\{u,v\} \in E\}$. The capacity of any edge incident with s or t is
equal to 1, and the capacities of the other edges are taken to be ∞.
Then the value of a maximum flow in $\hat{G}$ is equal to the size of a maximum
matching in G. This is seen by noting the following correspondence:

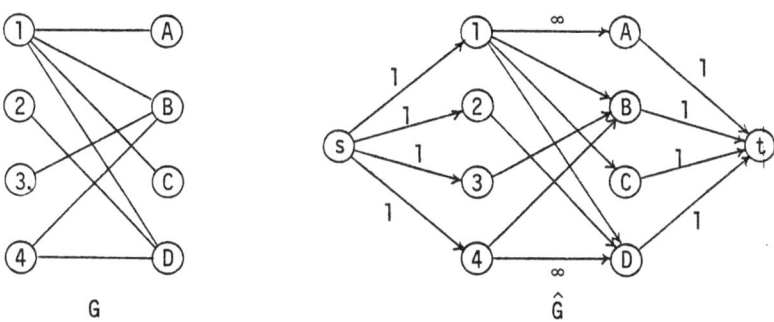

G $\hat{G}$

Conversion of a Bipartite Matching Problem
to a Flow Problem

a maximum flow will break up into disjoint s-t paths of length 3, each
carrying one unit. No two of these paths will start with the same edge
or end with the same edge. Hence, the edges from R to B in these
paths form a matching.

Exercise. Prove in detail that the value of a maximum flow in $\hat{G}$ is
equal to the size of a maximum matching in G.

     Similarly, a minimum cut in $\hat{G}$ corresponds to a minimum-cardinality
set of vertices of G containing a vertex incident with each edge of G.
     Hence the maximum size of a matching in a bipartite graph G is
equal to the minimum number of vertices needed to cover all the edges of G.

— 124 —

R. M. Karp

This theorem can be put in an attractive alternate form. Suppose we represent the bipartite graph G by a matrix with a row for each element of R, a column for each element of B, and an x in position (i,j) if {i,j} ε E. Then the restatement of the max-flow min-cut theorem in this case is:

König-Egervary Theorem. For any rectangular matrix containing x's in some of its positions, the maximum number of independent x's (no two in the same row or column) is equal to the minimum number of lines (rows or columns) needed to cover all the x's.

Computing Optimum Matchings

We next develop an approach to matching theory which applies to graphs which are not necessarily bipartite. A path is called simple if no two of its vertices are the same, except possibly the first and last. Given a matching M ⊆ E, call a vertex free if it is incident with no element of M. Call a path alternating if it contains no two consecutive elements of M, and no two consecutive elements of E-M. A simple alternating path between free vertices is called an augmenting path. Indicating the elements of M by wiggly edges (•∿∿•) and the elements of E-M by solid edges (•——•), we find that an augmenting path has the following appearance:

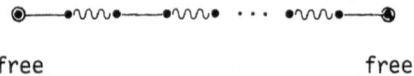

free   free

Theorem (Berge). There is an augmenting path relative to M ⟺ M is not a maximum matching.

Proof. $\Rightarrow$ Let P be an augmenting path relative to M. Then M' = M$\oplus$P is a matching, and has one more element than M has. [Here $\oplus$ denotes 'symmetric difference', so M' is obtained by deleting from M the elements in M$\cap$P, and adding the elements in P-M.]

$\Leftarrow$ Let N be a maximum matching. Then $|N| > |M|$. It follows that $|N| > |M|$. Consider the graph $G^* = (V, M\oplus N)$. In G^*, denote the edges in M-N as wiggly lines (•⌇⌇⌇•) and the edges in N-M as solid lines (•———•). Then G^* has more solid edges than wiggly edges. Also, since M and N are matchings, each vertex of G^* is incident with at most one wiggly edge and at most one solid edge. It follows that each component of G^* is either an isolated vertex, an alternating cycle (i.e., one consisting alternately of solid and wiggly edges) or a simple alternating path. One component must contain more solid edges than wiggly edges, and it can only be an alternating path beginning and ending with a solid edge. Such a path is an augmenting path relative to M. □

Exercise. Suppose we start with some initial matching and repeatedly augment along a shortest augmenting path. Let the successive paths so obtained be $P_1, P_2, \ldots, P_k, \ldots$. Then

   (a) length(P_k) $\leq$ length(P_{k+1}), k = 1,2,...,

   (b) no edge occurs in two augmenting paths of the same length.

This suggests that we can compute a maximum matching in a series of stages, in each of which we exhaust paths of a given length. In the bipartite case we can give an efficient implementation based on this idea; not surprisingly, the implementation is very similar to the one we used in the computation of maximum flows. The main difference is that, when we augment along a path, all the edges of the path are popped from the stack. Because of this feature, the number of PUSH and POP operations

during a single stage, in which we exhaust all the augmenting paths of a
given length, is $O(|E|)$.

We can also show that the number of stages will be only about $\sqrt{2\lceil V\rceil}$.
We can easily extend the proof of Berge's Theorem to yield the following
lemma.

Lemma. If M is a matching which is not maximum, and the cardinality of
a maximum matching is x, then there is an augmenting path relative to
M containing at most $\frac{|M|}{x-|M|}$ edges from M.

It follows from this Lemma by setting $|M| = x - \sqrt{x}$ that the first
$\sqrt{x}$ stages produce a matching of size $\geq x - \sqrt{x}$; hence at most $2\sqrt{x}$
stages are needed altogether. Since $x \leq \frac{|V|}{2}$, the number of stages
required is $\leq \sqrt{2\lceil V\rceil}$, and the running time is $O(|E|\sqrt{\lceil V\rceil})$. Tarjan has
given examples showing that this upper bound on the growth rate of the
running time for this algorithm is actually tight.

To complete our discussion of maximum matchings in bipartite graphs
(or, equivalently, maximum sets of independent x's in matrices), we show
how to construct a set of lines of minimum cardinality which cover all the
x's. Let us call the two types of vertices in a bipartite graph the boys
and the girls; these correspond to the rows and columns in the matrix.
Suppose we have constructed a maximum matching M. Consider the structure of the unsuccessful search for an augmenting path relative to M.
Let $\hat{G}$ be the set of girls reachable from some free boy by an alternating
path, and let $\hat{B}$ be the set of boys not reachable from any free boy by
an alternating path. Then $\hat{G}$ covers every x covered by a boy in $B-\hat{B}$
(where B is the set of all boys), so $\hat{G} \cup \hat{B}$ covers all the x's. Also,
$|\hat{G}| = |B-\hat{B}| -$ (number of free boys), whence $|\hat{G}\cup\hat{B}| = |B| -$ (number of

R. M. Karp

free boys) = $|M|$.

<u>Example.</u>

	A	B	C	D	E	F
1	x		x		x	⊗
2			x		x	
3			x		⊗	
4	x	x		⊗		
5			⊗			
6					x	

The free boys are 2 and 6. The search:

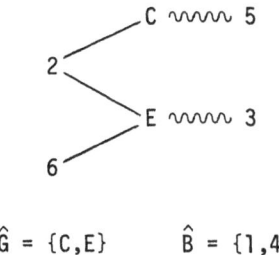

$$\hat{G} = \{C,E\} \qquad \hat{B} = \{1,4\}$$

The set of lines $\{1,4,C,E\}$ covers all the x's.

Now we apply augmenting path ideas to the solution of a frequently occurring optimization problem called the <u>assignment problem</u>. Suppose we have n men and n jobs, and have a cost c_{ij} of assigning man i to job j. We seek a 1-1 assignment of men to jobs which has minimum cost. In other words, we require that permutation π of $\{1,2,\ldots,n\}$ which minimizes $\sum c_{i,\pi(i)}$.

The first observation on the way to a solution method is that the optimum choice of a permutation is not affected by adding constants α_i to the rows and β_j to the columns of the cost matrix to produce a new cost matrix $(\bar{c}_{ij})$, where $\bar{c}_{ij} = c_{ij} + \alpha_i + \beta_j$. Given any permutation π

$$\sum \bar{c}_{i,\pi(i)} = \sum_{i=1}^{n} c_{i,\pi(i)} + \sum_{i=1}^{n} \alpha_i + \sum_{i=1}^{n} \beta_{\pi(i)} = \sum_{i=1}^{n} c_{i,\pi(i)} + \sum_{i=1}^{n} \alpha_i + \sum_{j=1}^{n} \beta_j .$$

Thus, the transformation adds the same constant to the cost of every permutation.

Secondly, suppose we succeed in transforming the cost matrix to a nonnegative matrix with a set of n independent zeros. Then we can use those zeros to form a permutation of cost 0, which is clearly optimum.

Given any nonnegative cost matrix, suppose that the maximum number of independent zeros is $k < n$. Then we can cover the zeros with k lines. Clearly, new zeros must be created among the cells that are not covered by any one of these k lines. If we subtract e from every row not in the cover, and add e to every column in the cover, then the net effect is to subtract e from every uncovered cell in the matrix, and to add e to every doubly covered cell. We choose e as the least value occurring in any uncovered cell.

Exercise. If S is a maximum set of independent zeros, then no element of S is doubly covered.

Example. Suppose the cost matrix is the one shown in (a) below. A convenient way to start is to subtract the minimum element in each row from all elements of the row, and then do the same to the columns. This produces the matrix shown in (b). The rest of the display shows the successive matrices reached in the computation.

R. M. Karp

$$
\begin{array}{c}
\begin{array}{cccc} A\ B\ C\ D \end{array}\\[2pt]
\begin{array}{c} 1 \\ 2 \\ 3 \\ 4 \end{array}
\left[\begin{array}{cccc}
5 & 2 & 2 & 1\\
1 & 6 & 7 & 4\\
1 & 6 & 5 & 3\\
1 & 8 & 9 & 7
\end{array}\right]
\qquad
\left[\begin{array}{cccc}
\cancel{①} & 0 & 0 & \\
④ & 4 & 5 & 3\\
0 & 5 & 5 & 3\\
0 & 6 & 7 & 7
\end{array}\right]
\qquad
\left[\begin{array}{cccc}
① & 0 & \\
① & 1 & 2 & \\
0 & 2 & 2 & ⓪\\
0 & 3 & 4 & 4
\end{array}\right]
\qquad
\left[\begin{array}{cccc}
8 & 0 & ⓪ & 1\\
0 & ⓪ & 1 & 0\\
0 & 1 & 1 & ⓪\\
⓪ & 2 & 3 & 4
\end{array}\right]
\end{array}
$$

$$
\qquad\qquad\qquad\ \ e=3 \qquad\qquad\ \ e=1
$$

(a)   (b)   (c)   (d)

In this example progress was misleadingly smooth, since the maximum
number of independent zeros increases at each iteration. This is not the
case in general. We give two proofs that the algorithm terminates.
The first proof is by noting that every iteration reduces the sum of the
entries in the matrix by an integer (assuming the matrix was originally
all-integer). The second proof stems from the fact that the set of
columns reachable from a free row by an alternating path, in which the
edges correspond to cells marked zero, strictly grows at each iteration.
The rows and columns reachable before the iteration step continue to be
reachable, and the new zero is so placed that it makes at least one more
column reachable. We illustrate by exhibiting the reachable rows and
columns in matrix (b), and then in matrix (c) (before the 3-D element is
circled).

(b)   (c)

Since we either augment the matching or reach a new column at every
iteration, an augmentation must occur every n iterations, and the entire
computation must terminate within n^2 iterations. If, for each row, we

use a priority queue to represent the elements of that row which are not
covered by a column in the minimum cover, then each iteration can be carried out in $O(n \log n)$ primitive steps (addition, subtraction, comparison),
and hence the assignment problem can be solved in $O(n^3 \log n)$ steps.

While the above algorithm (called the Hungarian method) is a nice
application of the König-Egervary theorem, it is not the most efficient
algorithm known for the solution of the assignment problem. We next give
another method, based on shortest-path calculations, which solves the problem in $O(n^3)$ steps. First we review a few facts about shortest paths.
Let $G = (V,E)$ be a digraph with $V = \{1,2,\ldots,n\}$, let each edge (i,j)
have weight $d_{ij} \geq 0$, and let π_i^* denote the minimum weight of a path
from 1 to i. Then, clearly, for each edge (i,j),

$$\pi_i^* + d_{ij} - \pi_j^* \geq 0 ,$$

and

$$\pi_i^* + d_{ij} - \pi_j^* = 0$$

if (i,j) lies in a minimum-weight path from 1 to j.

Now suppose we are given a nonnegative $n \times n$ matrix (c_{ij}) in which
a set of independent zeros is circled. We construct an associated digraph with $2n$ vertices, of which n correspond to the rows and n, to
the columns. If cell i-j is not circled, then we have an edge of
weight c_{ij} from row i to column j. If i-j is circled, then we have
an edge of weight 0 from j to i. In this digraph, we may use the
Dijkstra algorithm to compute shortest paths from the set of free rows to
all rows and columns. This requires a slight modification, since the
original algorithm is given for a single source, and now every one of the
free rows is treated as a source. Let γ_i denote the weight of a

shortest path to row i, and let δ_j denote the weight of a shortest path to column j. Now consider the matrix $(\bar{c}_{ij}) = (c_{ij} + \gamma_i - \delta_j)$. By the foregoing remarks about shortest-path problems, this will be a non-negative matrix, and all the cells that lie in shortest paths will become 0. Also, the cells that were circled will remain at 0. Every entry in a shortest path to a free column becomes 0, so every such path may be used as an augmenting path, and thus the size of the matching (i.e., the set of circled independent zeros) can be increased by at least 1. Thus, after n iterations, we obtain n independent zeros, and the assignment problem is solved. Each shortest-path calculation takes $O(n^2)$ steps, and thus the entire process is completed in $O(n^3)$ steps.

We give a computational example. At each step each row is labelled with γ_i and each column with δ_j. A matching is indicated in each matrix, and the augmenting path used to get that matching from the previous one is shown.

```
        1 2 2 1              0 4 3 2              0 0 1 1

   0  ( 5 2 2 1 )       4  ( 4 ⓪ 0 0 )      0  ( 8 ⓪ 1 2 )
   0  | 1 6 7 4 |       0  | ⓪ 4 5 3 |      0  | ⓪ 0 2 1 |
   0  | 1 6 5 3 |       0  | 0 4 3 2 |      1  | 0 0 ⓪ 0 |
   0  ( 1 8 9 7 )       0  ( 0 6 7 6 )      0  ( 0 2 4 2 )

         (a)                  (b)                  (c)
```

```
        ( 8   0---ⓞ   1 )
        | 0---ⓞ  1    0 |
        | 1    1  0---ⓞ |
        ( ⓞ   2  3    1 )

                (d)
```

R. M. Karp

There is a heuristic probabilistic argument which indicates that, on the average, this process should terminate within $O(\log n)$ iterations. The set of shortest paths consists of disjoint trees rooted at the free rows. Each free column is in exactly one of these trees. If there are x free rows and x free columns, and the tree in which a given free column occurs is chosen at random, then the probability that at least one free column occurs in the tree rooted at a given free row is $1 - (1 - \frac{1}{x})^x \sim 1 - e^{-1}$. Thus the expected number of augmentations possible is about $(1-e^{-1})x$. If a fixed fraction of the free rows becomes matched at each iteration, then the number of iterations needed is about $\log n$. Unfortunately no rigorous probabilistic analysis is available to support the heuristic argument, which suggests an expected running time of $O(n^2 \log n)$.

We now present Edmonds' algorithm for constructing maximum matchings in general graphs. Given a graph $G = (V,E)$, and a matching $M \subseteq E$, the algorithm either finds an augmenting path relative to M or establishes that there is none. The algorithm starts with some free vertex v (if no free vertex exists then M is maximum) and builds a search tree T rooted at v, such that every path directed away from v is alternating. The vertex v is designated as "outer", and each vertex in T is classified as outer or inner in such a way that outer and inner vertices alternate along any path of T directed away from v. The general step in the construction is:

   (1) Choose an outer vertex $w \in T$, and some edge $\{w,x\} \in E-M$ which has not been explored before. If no such edge exists, go to H.

   (2) If x is free, go to A.

(3) If x is matched to some vertex y, add edges (w,x) and
     (x,y) to T, designate x as inner and y as outer.

(4) If x is inner, go to (1).

(5) If x is outer, go to B.

To complete the discussion, we must explain what happens at exits B,
H and A.

Exit B. We have found an edge between two outer vertices, yielding a
situation of the following type.

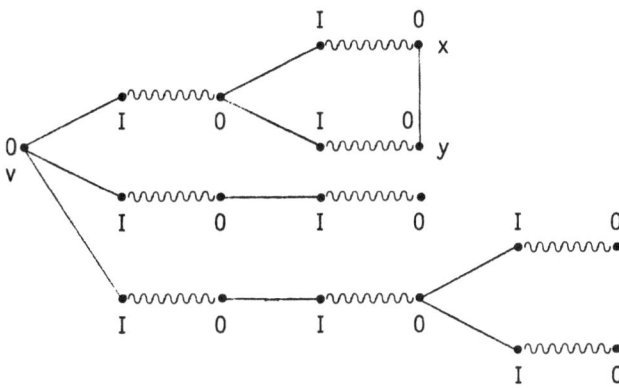

A cycle of odd length has been formed. In this case we "shrink" the
cycle by contracting all its vertices into a macro-vertex β. β is adjacent to any of the vertices in G to which one of the vertices in the
cycle is adjacent. β is regarded as a free vertex if and only if it
contains the root v. β replaces w in the search tree as an outer
vertex, and the search continues. An odd cycle such as β is called a
"blossom".

Exit H. In this case it is impossible to extend the tree T. If we look
at T together with all edges incident with any vertex of T, we get a

structure like the following one:

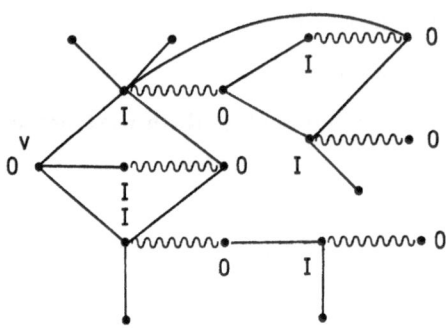

We note that

(1) The outer vertices may be macro-vertices obtained by shrinking blossoms, but the inner vertices are ordinary vertices.

(2) The only free vertex is v.

(3) Edges joining vertices in T to vertices not in T are solid, and join T at inner vertices.

Hence we assert that no vertex in T can occur in an augmenting path. Since T contains only one free vertex, we may assume that such a path enters T along a solid edge at an inner vertex. It then must alternately visit outer vertices along wiggly edges and inner vertices along solid edges. Thus the path is "trapped" -- it can neither reach v nor leave T.

In this case we delete T from the graph, together with all edges incident with vertices of T. T will have to be restored if the matching M is later augmented.

Exit A. In this case we have found a simple alternating path between free vertices, and expect to augment the matching. However, we must take account of the fact that some of the outer vertices in the path may be macro-vertices. This situation is easy to handle. Suppose the vertex marked 0 is a macro-vertex obtained by shrinking some odd cycle.

Unshrinking that cycle, we obtain:

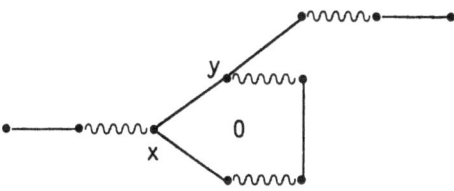

But we may interpolate a part of the cycle of even length into the path; for example,

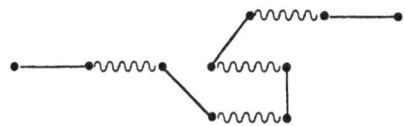

Continuing to "unshrink" in this way, we eventually obtain an augmenting path relative to M in the original graph.

We illustrate the algorithm using the following example.

R. M. Karp

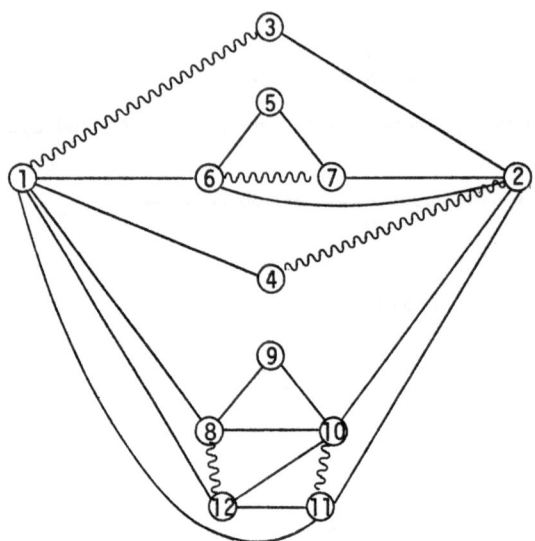

The free vertices are 5 and 9. Starting the search from vertex 5, we get:

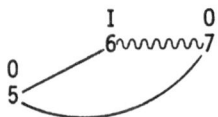

Shrinking the blossom 5 6 7 to obtain a macrovertex and continuing

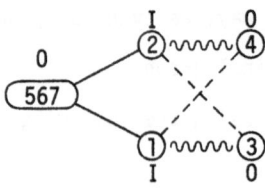

Exit H is reached. Vertices 1, 2, 3, 4, 5, 6 and 7 are deleted, leaving

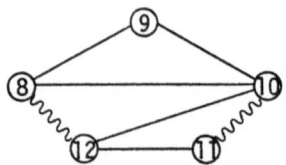

R. M. Karp

Now there is only one free vertex, so clearly no augmentation is possible, The given matching is maximum.

Even and Kariv [22] have designed an algorithm similar to Edmonds', but organized in stages, like our bipartite matching algorithms. With careful implementation, their method runs in $O(|V|^{5/2})$ steps.

6.   On the Computational Complexity of Combinatorial Problems[†]

Abstract: A large class of classical combinatorial problems, including most of the difficult problems in the literature of network flows and computational graph theory, are shown to be equivalent, in the sense that either all or none of them can be solved in polynomial time. Moreover, they are solvable in polynomial time if and only if every problem solvable by a polynomial-depth backtrack search is also solvable by a polynomial-time algorithm. The technique of deriving these results through problem transformations is explained, and some comments are made as to the probable effect of these results on research in the field of combinatorial algorithms.

Nondeterministic Algorithms, Backtrack and Exponential Growth

An example: 3-Coloring a Graph

Let us call a graph G 3-colorable if its nodes can be partitioned into three color classes C_1, C_2 and C_3, such that no two adjacent nodes lie in the same color class. While there is no known efficient algorithm to decide in general whether a graph is 3-colorable, one can

[†]Reprinted by permission from Networks 5 (1975), 45-68.

write a concise "nondeterministic algorithm" for this coloring problem as follows. Define a _partial_ _coloring_ as an assignment of some of the nodes to the three color classes, such that no two adjacent nodes lie in the same color class. The unassigned nodes are called _free_, and a free node is _eligible_ _for_ C_i if none of its neighbors are in C_i. The nondeterministic algorithm proceeds by building up larger and larger partial colorings, making forced choices when they exist, halting when a dead end is reached, and otherwise making an arbitrary choice for the color of some free node.

Conventions. G is connected, and its nodes are named $v_1, v_2, \ldots, v_n$. For $i = 1, 2, 3$, V_j is a variable denoting the set of free nodes eligible for exactly j color classes.

```
        Start
        Assign v₁ to C₁
        Assign the lowest numbered neighbor of v₁ to C₂
        Comment: There is no loss of generality in fixing the colors of
                 two adjacent nodes.
     Do 1 while at least one free node remains
     If V₀ ≠ φ, Halt
        Comment: A dead end is reached.
        Else, if V₁ ≠ φ then assign each node in V₁ to the unique color
            class for which it is eligible.
        Comment: These assignments are forced.
        Else, let vₖ be the lowest numbered node in V₂.
        Choose one of the color classes for which vₖ is eligible, and
                assign vₖ to that color class.
        Comment: It is the arbitrary choice at this stage which makes
                 the algorithm nondeterministic.
        1. Continue.
    Nondeterministic Algorithm for the 3-Coloring Problem
```

R. M. Karp

The nondeterministic algorithm can follow many possible computational paths for a single graph G, depending on the sequence of arbitrary choices made at step 1; it is easy to construct a family of examples for which the number of such paths grow exponentially with n. But each particular computational path terminates within n iterations of the Do-loop, either by reaching a dead end or by generating a 3-coloring. Thus, we may regard the nondeterministic algorithm as running in time proportional to n. Of course, this does not mean that one could program a computer to execute this 3-coloring algorithm in time proportional to n. To do so, one appears to require either an oracle which will infallibly indicate the correct branch to take at each choice point, or an unbounded capacity for parallel computation, in order to follow all the computation paths concurrently. One may confidently state that neither of these facilities is in prospect of becoming available.

There is a natural way to convert the above algorithm into a deterministic one. Whenever a choice is called for between two alternatives, each leading to a partial coloring, both partial colorings are generated. One of them is placed on a pushdown stack for later reference, and computation proceeds on the other one. Whenever a dead end is reached, a partial coloring is popped from the pushdown stack, and computation continues. The process ends either when a 3-coloring is found, or when the pushdown stack is emptied, indicating that all computation paths have been explored. By this technique, the nondeterministic algorithm is converted to a backtrack algorithm. Unfortunately, the execution time of the backtrack algorithm is determined not by the length of a single computation path, but by the total number of partial colorings generated along all computation paths -- and this number can grow as 2^n.

R. M. Karp

There is a wide class of combinatorial problems that can be tackled by the trial-and-error approach embodied in the backtrack method and its close relative, the branch-and-bound method. While these methods are useful and general, their effectiveness is limited by the typically exponential growth of their execution time as a function of problem size.

Sometimes it is possible to do better. For example, we can set up an algorithm, similar to the one above, to decide whether a connected graph is 2-colorable. It is easy to see that, in this case, the algorithm will proceed towards either a 2-coloring or a dead end without making any arbitrary choices. Thus only one computation path will be followed, and the total number of partial colorings generated will be less than or equal to n.

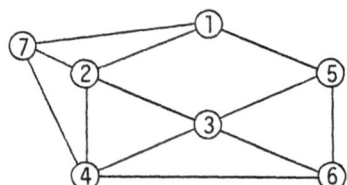

Fig. 1 A Graph G that is not 3-Colorable

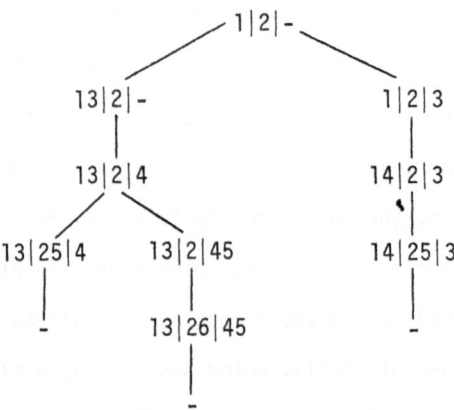

Fig. 2 A Tree of Computation Paths

R. M. Karp

A less trivial example is the matching problem in graphs. A matching is a subset M of the set of arcs of G, such that each node of G meets exactly one arc of M. An efficient algorithm to find a matching of maximum cardinality (or, more briefly, a maximum matching) has been disco⁻ vered by Edmonds [16]; it depends on a theorem of Berge [3]. Given a matching M, let us call a node free if it is incident with no arc in M. Call a path in G alternating if its arcs are alternately in M and not in M. Define an augmenting path as a simple (i.e., without repeated nodes) alternating path between free nodes. Berge's theorem states that a matching M is maximum if and only if it does not admit an augmenting path. Moreover, when an augmenting path does exist, one can increase the cardinality of the matching by reversing the polarity of each arc in the augmenting path; those in the matching are dropped, and those not in the matching are added. Thus, one can obtain a maximum matching by repeatedly searching for augmenting paths, which yield matchings of greater and greater cardinality; the process terminates when augmenting paths no longer exist.

To implement this approach, it is necessary to have an efficient method of finding an augmenting path or determining that none exist. This problem turns out to be extremely tricky; the first satisfactory solution was given by Edmonds [16].

With careful implementation, the augmenting path approach yields an algorithm to find maximum matchings in a number of computation steps pro- portional to $n^{5/2}$, where n is the number of nodes [22]. When the graph has no cycles of odd length the search for augmenting paths is espe⁻ cially simple [32]. Such time bounds are, of course, far

R. M. Karp

better than the exponential time bounds associated with a straightforward
backtrack, or branch-and-bound, approach.

There are many other examples where the augmenting path approach
yields algorithms whose execution time is polynomial, rather than exponential, in the size of the input. Examples are the maximum flow problem,
and the matroid intersection and partition problems. An excellent survey
of augmenting path theory is given in the forthcoming book by Lawler [48].

In general, the problems studied in the field of combinatorial algorithms can be fruitfully divided into two classes. In the first class are
those problems which, like the matching problem, can be solved in polynomial time. Usually problems fall into this class only after some signifi-
cant insight into their structure has been gained. In the second class
are those problems for which no solution method has been found which, in
its worst case, is decisively better than the "general-purpose" backtrack
method; algorithms for such problems cannot be guaranteed to run in less
than exponential time. Important examples of problems in the first class
are the shortest path, maximum flow, assignment and transportation problems, as well as the minimum spanning tree problem, the problem of testing whether a graph is planar, and a number of sequencing problems endowed
with special structure. The second class includes such old chestnuts as
0-1 integer programming, the travelling salesman problem, the graph isomorphism problem, and a host of coloring, covering, packing and partition-
ing problems associated with networks. Alan Cobham, Jack Edmonds, and
Michael Rabin appear to have been the first to stress the distinction
between these two classes of problems.

The present paper is an account of a body of recent research which
shows that most of the well-known problems which appear to be intrinsically

exponential are equivalent, in the sense that either all of them or none of them admit of polynomial-time algorithms. Moreover, there is strong evidence that all these equivalent problems are in fact worse than poly-nomial-bounded; for, if they did admit of polynomial-time algorithms, then there would follow the highly implausible result that whenever any problem can be solved by a backtrack algorithm whose computation paths are of polynomial-bounded length, that problem can also be solved by a determin-istic algorithm operating within a polynomial time bound.

The Class P

In stating formal results about the equivalence of computational pro-blems, we will restrict attention to decision problems, rather than pro-blems which require the computation of an optimum value. This convention will require us to resort to artifice at times. For example, it is natural to view the travelling salesman problem as that of determining the cost of a minimum tour, given a network with nonnegative integer costs on its arcs. But, to conform with our convention, we will instead regard the travelling salesman problem as having two parts to its input: a network, and a conjectured value for the optimum tour. The decision problem, then, is to determine whether there exists a tour of value less than the conjec-tured optimum. It should be clear, however, that the "natural" version of the travelling salesman problem can be solved in polynomial time if and only if the associated decision problem can be solved in polynomial time.

The inputs to our decision problems will be structures such as graphs, networks, vectors of integers, Boolean formulas, etc. The second important convention required for the formal development is that the inputs are somehow encoded as sequences of 0's and 1's; this is not unrea-sonable, since such an encoding is implicit in the preparation of the

R. M. Karp

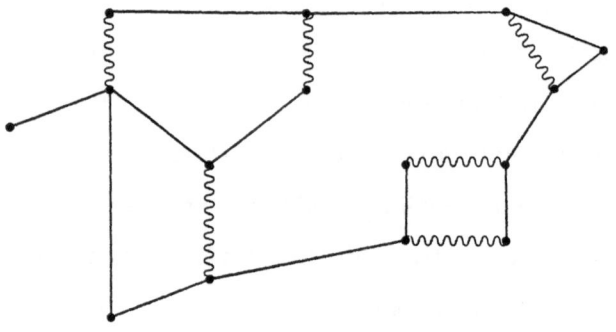

(a) A Matching

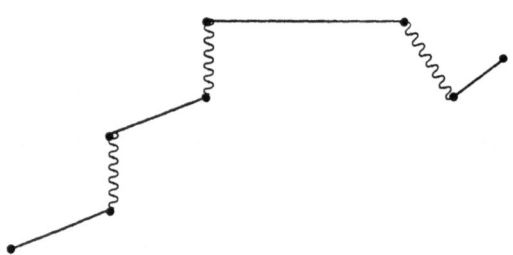

(b) An Augmenting Path

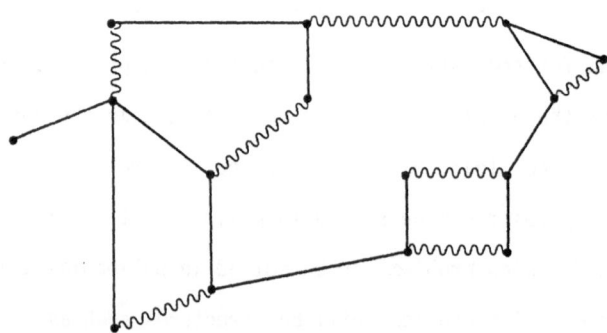

(c) A Maximum Matching

Fig. 3 The Process of Augmenting a Matching

R. M. Karp

input data for any digital computer. Usually it will be unnecessary to consider the precise details of the digital encoding of the input.

With our two conventions, any decision problem may be equated with a subset L of $\{0,1\}^*$, the set of all finite strings of zeros and ones. L is just the set of encoded inputs for which the correct output is "yes."

We define P, the set of all decision problems solvable by a deterministic algorithm in polynomial time, as follows. The set of strings L is in P if there is an algorithm A such that:

(i) A gives the output "yes" for input strings in L and "no" for input strings in $\{0,1\}^* - L$;

(ii) A operates in polynomial time; that is, there exists a polynomial $f(\cdot)$ such that, for every input string x, the number of steps required by A in processing x is less than or equal to $f(|x|)$, where $|x|$ denotes the length of x. The algorithm A is said to <u>recognize</u> the set L, or to <u>accept</u> the strings in L.

The definition of P is really not completed until we define what is meant by an algorithm, and what is meant by a step. It is convenient to develop these formal definitions in terms of an abstract machine called a Turing machine [35], but it is important to realize that there is considerable latitude in the form of these definitions. There are many plausible formal models of a digital computer to be found in the computational complexity literature, and all of them yield exactly the same definition of P (we rule out as implausibly powerful certain abstract machines that are capable of unbounded parallelism or of multiplying arbitrarily long numbers in a single step; and we consider as implausibly weak "counter

R. M. Karp

machines" which can represent numbers in unary notation, but not in binary). It will suffice for our purposes to regard P as the class of problems solvable in polynomial time by a digital computer, idealized so that it has unbounded backup storage, and hence can accept arbitrarily long input strings.

To clarify the definition of P, let us consider the 0-1 knapsack problem: given positive integer vectors $(A_1, A_2, \ldots, A_n)$ and $(C_1, C_2, \ldots, C_n)$, and a positive integer d, find a 0-1 vector $(x_1, x_2, \ldots, x_n)$ which maximizes $\sum_{j=1}^{n} A_j x_j$, subject to $\sum_{n=1}^{n} C_j x_j \leq d$. To state the problem as a decision problem, we take as an additional input a positive integer value b for $\sum A_j x_j$; and ask whether there is a 0-1 solution to:

$$\sum_{j=1}^{n} A_j x_j \geq b$$

$$\sum_{j=1}^{n} C_j x_j \leq d .$$

A well-known dynamic programming algorithm for this problem goes as follows:

for $m = 1, 2, \ldots, n$ and $e = 1, 2, \ldots, d$ define $F(m,e)$ as the maximum value of

$$\sum_{k=1}^{m} A_k x_k \quad \text{subject to} \quad \sum_{k=1}^{m} C_k x_k \leq e ,$$

where the x_k are 0-1 variables. Then $F(m,e)$ satisfies the initial condition:

$$F(1,e) = \begin{cases} 0 , & e < A_1 \\ C_1 , & e \geq A_1 \end{cases}$$

and the recurrence $F(m,e) = \max\{F(m-1,e),F(m-1,e-C_m) + A_m\}$, $m = 2,\ldots,n$. The decision problem can be solved by tabulating the function $F(m,e)$, and then testing whether $F(n,d) \geq b$. The algorithm requires $O(nd)$ additions and comparisons; each of these operations can be carried out in a number of elementary logical operations on binary digits proportional to the lengths of the operands. To determine whether this dynamic programming algorithm is polynomial bounded, suppose that each of the A_j and C_j is an x-digit binary number; i.e., $1 \leq A_j \leq 2^x - 1$, and that $d \sim n(2^x-1)$, so that d can be expressed in about $x + \log_2 n$ digits. Thus, with reasonable encoding, the entire length of the input string is $O(nx)$, but the number of computation steps is $O(n^2 x \cdot 2^x)$, which, as x grows, is not bounded by a polynomial in nx; thus the algorithm is not polynomial bounded. On the other hand, if we put a uniform upper bound or the lengths of the binary numbers C_j and A_j, rejecting all inputs exceeding this upper bound, then the algorithm would run in polynomial time. Alternatively, if the input encoding represented the A_j and C_j as unary numbers, of length about 2^x, then the running time would be bounded by a polynomial in the length of the input string; but such padding of the input has to be rejected as unreasonable.

Problem Transformations

It is a standard technique in the field of mathematical programming to reduce one problem to another by giving a transformation which maps any input to the first problem into an equivalent input to the second problem. As early as 1960 Dantzig showed that a variety of combinatorial problems can be transformed into zero-one integer programming problems [12]. There are many examples of problem transformations in the field of network flows; two typical ones are the transformation which takes minimum-cost

R. M. Karp

circulation problems into Hitchcock transportation problems, and the transformation of Hitchcock transportation problems into assignment problems
[25]. Other well-known examples are the transformation of 0-1 integer
programs into 0-1 knapsack problems [5], and the transformation of the
maximum degree constrained subgraph problem into a matching problem [18].

Our results are derived from a systematic exploitation of such transformations. We restrict attention to transformations which are carried
out in polynomial time; this restriction excludes several of the foregoing examples.

Given two decision problems $L \subseteq \{0,1\}^*$ and $M \subseteq \{0,1\}^*$, we say
that $L \alpha M$ (L is transformable into M) if there is a function f such
that

(i) f maps strings into strings (i.e., f has $\{0,1\}^*$ as both its
        domain and range);

(ii) there is an algorithm to compute f in polynomial-time; and

(iii) the transformation f preserves the problem (i.e., $f(x) \in M$
        if and only if $x \in L$).

Thus, to decide if $x \in L$, one can compute $f(x)$, and then test whether
$f(x) \in M$. It follows that, if $L \alpha M$ and $M \in P$, then $L \in P$.

Later we will see examples of sets of problems, each of which is
transformable to any of the others; it follows that either all or none of
the problems in such a set are solvable in polynomial time.

To give a feeling for the crucial concept of transformability, we
introduce two particular problems -- the satisfiability problem and the
discrete multicommodity flow problem -- and sketch Donald Knuth's result
that satisfiability α discrete multicommodity flow [45].

The satisfiability problem arises in elementary mathematical logic.

R. M. Karp

Let A,B,C,... denote propositional variables which may be either true or false. Let $\bar{A},\bar{B},\bar{C},...$ denote the negations of these variables. Then $\bar{A}$ is true if and only if A is false. Let $A,\bar{A},B,\bar{B},...$ be called <u>literals</u>. Let $\cup$ denote the 'or' connective, and let $\cap$ denote the 'and' connective. A <u>clause</u> is simply the 'or' of a set of literals; e.g., $A \cup \bar{B} \cup \bar{D}$. A <u>conjunctive</u> <u>normal</u> <u>form</u> <u>expression</u> is the 'and' of a set of clauses; e.g. $(A \cup \bar{B} \cup \bar{D}) \cap (\bar{A} \cup \bar{C}) \cap (\bar{B} \cup C) \cap D$. If we assign each variable the value T (true) or F (false), then the entire expression assumes the value T or F. For example, taking $A = D = T$, $B = C = F$, the evaluation is

$$(T \cup T \cup F) \cap (F \cup F) \cap (T \cup F) \cap T = T \cap F \cap T \cap T$$

$$= F .$$

A conjunctive normal form expression is called <u>satisfiable</u> if some assignment of truth values to the variables makes the expression evaluate to T. The satisfiability problem is to determine, of an arbitrary conjunctive normal form expression given as input, whether that expression is satisfiable.

The discrete multicommodity flow problem is a simplified version of a problem well-known to readers of this journal. Given an undirected graph and a collection of disjoint source-sink pairs $(s_1,t_1),(s_2,t_2),...,$ (s_k,t_k), it asks whether there exists a set of k node-disjoint paths, the ith of which joins s_i with t_i. Figure 4 gives an example with $k = 3$ in which the required paths do exist.

We give a construction which, in polynomial time, transforms any instance of the satisfiability problem into an equivalent instance of the discrete multicommodity flow problem. The source-sink pairs in the

R. M. Karp

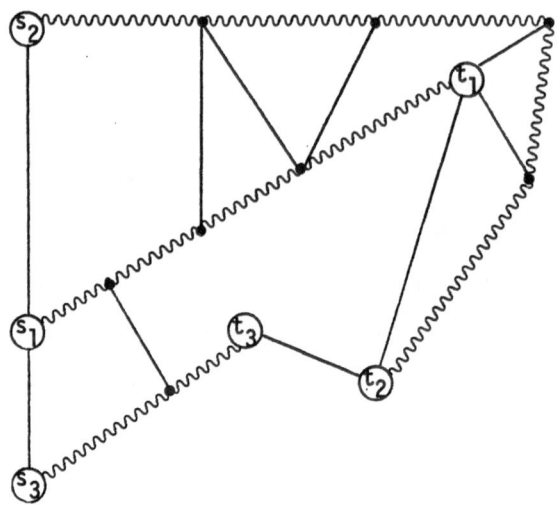

**Fig. 4 A Feasible Instance of the Discrete
Multicommodity Flow Problem**

constructed flow problem will be in one-to-one correspondence with the
clauses in the given satisfiability problem -- pair (s_i, t_i) will correspond to the clause C_i. The graph will be obtained by joining together a
number of subgraphs, each corresponding to a variable. If variable A
occurs in clauses $i_1, i_2, \ldots, i_p$, and $\bar{A}$ occurs in $j_1, j_2, \ldots, j_q$, then
the subgraph corresponding to A looks as follows:

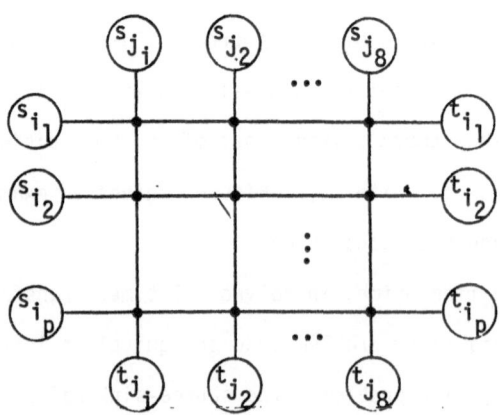

R. M. Karp

The subgraph is designed so that, within it, it is possible to join all the s_{i_ℓ}'s to the corresponding t_{i_ℓ}'s by node-disjoint horizontal paths, or to join all the s_{j_m}'s to the corresponding t_{j_m}'s by node-disjoint vertical paths, but every $s_{i_\ell} - t_{i_\ell}$ path has a node in common with every $s_{j_m} - t_{j_m}$ path. The overall graph is obtained simply by identifying, as a single node, all the occurrences of each s_i (or t_i) in the various subgraphs.

It is easy to see that, if the given conjunctive normal form expression is satisfiable, then the associated discrete multicommodity flow problem is feasible. Given an assignment of truth values satisfying the expression, we obtain the required set of node-disjoint paths as follows: if variable x is assigned the value 'true,' select the horizontal paths in the subgraph for x; if x is 'false,' select the vertical paths. Thus, if either x or $\bar{x}$ occurs in clause C_i and is true in the assignment, then s_i and t_i will be joined in the subgraph for x.

To complete the proof it is necessary to check that feasibility of the derived network problem implies satisfiability in the original problem, and that the transformation can be performed in polynomial time. This is easily done; hence, satisfiability α discrete multicommodity flow. It follows that a polynomial-time decision method for the discrete multicommodity flow problem would yield a polynomial-time algorithm for the satisfiability problem.

It is also not difficult to show that discrete multicommodity flow α satisfiability. Thus we have an equivalence between a network flow problem and a problem in elementary logic! If one of the two is solvable in polynomial time, then so is the other.

R. M. Karp

NP, Cook's Theorem and NP Completeness

In this section we introduce a class of decision problems called NP, and state a key theorem due to Stephen Cook, which asserts that every problem in NP is transformable to the satisfiability problem. NP includes nearly all the old chestnuts of the combinatorial optimization literature. Hence, if anyone finds a polynomial-time algorithm for the satisfiability problem, he will _ipso facto_ have found polynomial-time algorithms for a host of classical unsolved algorithmic problems. Viewing Cook's result more soberly, one may suspect that the satisfiability problem is not in P, since the consequences of a polynomial-time algorithm for its solution would be so radical.

NP is the class of decision problems solvable by nondeterministic algorithms operating in polynomial time. Informally, a nondeterministic algorithm is one which, in addition to all the usual features of algorithms, is capable of making an arbitrary choice between two alternative directions in which to branch. An input is accepted if some sequence of arbitrary choices yields a computation sequence which halts with the output "yes." A nondeterministic algorithm A is said to operate in polynomial time if there is a polynomial $f(\cdot)$ such that, on any input x, the length of every computation sequence is bounded by $f(|x|)$. As in the case of deterministic algorithms, it is convenient to frame the formal definition of NP in terms of Turing machines, but all plausible definitions of a nondeterministic polynomial-time algorithm would lead to the same class NP.

Since the concept of a nondeterministic polynomial-time algorithm may seem exotic, we give three alternate informal criteria for a set of strings L to be in NP. As usual, we think of strings as the encodings of

R. M. Karp

integers, graphs, networks or other structures.

Backtrack Criterion: There is a backtrack search algorithm of polynomial-bounded depth which accepts x if an only if x ε L.

Existence Criterion: Membership of x in L is equivalent to the existence of an appropriate structure y associated with x. For example, x may be the encoding of a graph, and y, the encoding of a Hamilton circuit, a 3-coloring, a clique of a given size, etc. Or x may be the encoding of a system of linear inequalities with integer coefficients, and y, the encoding of a 0-1 vector giving a feasible solution. It is required that

  (a) the length of y (the encoding of the desired structure) is bounded by a polynomial in the length of x, and

  (b) there is a simple (i.e., deterministic polynomial-time) algorithm to check, given the pair (x,y), whether y is an appropriate structure associated with x.

Proof Criterion: There is a consistent finite system S of axioms and rules of inference, and a polynomial f, such that x ε L if and only if the statement 'x ε L' is provable within S by a proof of length ≤ f(|x|).

The nondeterministic algorithm given in the Nondeterministic Algorithms section establishes that the 3-coloring problem is in NP. The problem of deciding whether a graph G has a Hamilton circuit (i.e., a cycle passing through each node exactly once) is easily seen to be in NP, by either the backtrack criterion or the existence criterion. Similarly the satisfiability problem and the problem of existence of a 0-1 solution to a system of linear inequalities with integer coefficients are both readily seen to be in NP.

Cook's Theorem [8]: Every problem in NP is transformable to the satisfiability problem.

Cook proves this theorem by specifying a single "master transformation" $F(A,f,x)$, where A is a nondeterministic algorithm (more technically, a nondeterministic one-tape Turing machine), f is a polynomial such that A operates within the time bound $f(|x|)$, and x is a string. The image of the transformation is a string which is the encoding of a conjunctive normal form expression. Cook shows that the transformation $F(A,f,x)$ can be computed deterministically in polynomial time, and that the expression $F(A,f,x)$ is satisfiable if and only if x is accepted by algorithm A. Thus, given a set of strings L in NP, there must be a polynomial-time nondeterministic algorithm A_L, operating in polynomial time f_L, that accepts L. Then L is transformable to satisfiability via the transformation $f(x) = F(A_L, f_L, x)$.

It follows that, if satisfiability could be tested in polynomial time, then every problem in NP would also be in P. On the other hand, it is trivially true that $P \subseteq NP$, since deterministic algorithms are a special case of nondeterministic ones. Summing up, we have the following corollary.

Corollary. $P = NP$ if and only if Satisfiability ε P.

Cook's Theorem shows that Satisfiability is, in a sense, as hard as any problem in NP. It turns out that many of the unsolved problems in the field of combinatorial algorithms share this distinction. Call a problem

L NP-complete if

    (i) $L \in$ NP

    (ii) Satisfiability α L.

    If L is NP-complete then, by Cook's Theorem, L α Satisfiability; thus, $L \in$ P if and only if Satisfiability $\in$ P, and we may make the following claim.

    Either all NP-complete problems are in P, or none are. The former is true if and only if P = NP.

A Sampling of NP-complete Problems

    During the past few years there has been intensive activity aimed at showing that particular problems are NP-complete. Whenever a problem is found to be NP-complete, we know that it is not isolated or unique, but instead is linked to a large number of classical hard problems. Such a discovery should lead all but the most confident and optimistic of us to despair of obtaining a polynomial-time algorithm for such a problem. Reasonable measures to take when confronted with such a problem are discussed in the next subsection.

    The usual method of showing that L is NP-complete is to check (usually by inspection) that $L \in$ NP, and then to show that Satisfiability α L. In showing that Satisfiability α L, we may make use of the transitivity of the relation α, in view of which it suffices to show that any NP-complete problem is transformable to L. The more specialized a problem appears, the more interesting is its NP-completeness. Thus we do not bother to include in the following list such problems as the integer programming problem, the quadratic assignment problem, the general set covering problem, the fixed charge transportation problem and the knapsack problem, preferring instead to list special cases of these problems. Most

R. M. Karp

of the problems in the original list I published in 1972 [41] are super-
seded here by more specialized problems which have since been discovered
to be NP-complete. The following list is not intended to be exhaustive.
The criteria for selecting a problem for inclusion were simplicity, sur-
prise value and direct relevance for this readership. Because of the last
criterion, several interesting examples related to automata and language
theory, n-person games, computer code optimization, etc. are omitted.
Each problem is specified by stating its input, and then the property that
must hold in order for the input to be accepted.

A List of NP-Complete Problems

1. Satisfiability with Three Literals per Clause [8]

   Input: A conjunctive normal form expression with at most three
   literals per clause.

   Property: The expression is satisfiable.

2. Planar Degree Constrained 3-Coloring [30]

   Input: A planar graph in which each node is of degree less than or
   equal to 4.

   Property: The graph is 3-colorable.

3. Clique [8], [53]

   Input: A graph G and a positive integer k.

   Property: G has a clique (complete subgraph) with k nodes.

4. Degree Constrained Node Cover [30]

   Input: A graph G in which the maximum node degree is 3, and a
   positive integer k.

   Property: G has a set S of k nodes such that every node not in
   S is adjacent to a node in S.

5. Degree Constrained Planar Node Cover [30]

Input: A planar graph in which the maximum node degree is less than
or equal to 6, and a positive integer k.

Property. G has a set S of k nodes such that every node not in
S is adjacent to a node in S.

6. Exact Cover [41]

Input: A collection $\{S_1, S_2, \ldots, S_n\}$ of finite sets.

Property: There is a subcollection $\{S_{i_1}, S_{i_2}, \ldots, S_{i_k}\}$ of disjoint
sets such that $\bigcup_{\ell=1}^{k} S_{i_\ell} = \bigcup_{j=1}^{n} S_j$.

7. Hitting Set [41]

Input: A collection $\{S_1, S_2, \ldots, S_n\}$ of finite sets.

Property: There is a set T such that, for i = 1,2,...,n,
$|T \cap S_i| = 1$.

8. 3-Dimensional Matching [41]

Input: A $p \times p \times p$ array, certain positions of which are specified
as "occupied."

Property: There is a set of p occupied positions, no two of which
agree in any of their three coordinates.

9. Cubic Hamilton Circuit [30]

Input: A cubic graph G.

Property: G has a Hamilton circuit.

10. Directed Hamilton Circuit [41]

Input: A directed graph G.

Property: G has a Hamilton circuit.

11. Degree Constrained Directed Planar Hamilton Line [30]

Input: A directed planar graph G in which each node has out-degree
≤ 4 and in-degree ≤ 3.

12. Feedback Node Set [41]

Input: A directed graph G and a positive integer k.

Property: G has a set of k nodes whose removal breaks all directed cycles.

13. Feedback Arc Set [41]

Input: A directed graph G and a positive integer k.

Property: G has a set of k arcs whose removal breaks all directed cycles.

14. Graph Partition [30], [54]

Input: A graph G with 2p nodes and a positive integer k.

Property: The nodes of G can be partitioned into two p-node sets S_1 and S_2, such that the number of arcs running between S_1 and S_2 is less than or equal to k.

15. Simple Maximum Cut [30]

Input: A graph G and a positive integer k.

Property: The nodes of G can be partitioned into two sets S_1 and S_2, such that the number of arcs running between S_1 and S_2 is at least k.

16. Optimal Linear Arrangement [30]

Input: A graph G and a positive integer k.

Property: There is a one-to-one function Π mapping the nodes of G onto a set of consecutive integers such that the sum of $|\Pi(u) - \Pi(v)|$, over all arcs (u,v), is less than or equal to k.

17. Numerical Partition [41]

Input: A set of positive integers $\{A_1, A_2, \ldots, A_n\}$.

Property: There is a set $S \subseteq \{1,2,\ldots,n\}$ such that $\sum_{i \in S} A_i = \sum_{i \notin S} A_i$.

R. M. Karp

18. Partition into Triplets [29]

Input: A set of positive integers $\{A_1,A_2,\ldots,A_n\}$, and positive

integers B and D.

Property: The index set $\{1,2,\ldots,n\}$ can be partitioned into sets

$S_1,S_2,\ldots,S_p$ such that $p \leq D$ and, for $j = 1,2,\ldots,p$,

$|S_j| \leq 3$ and $\sum_{i \in S_j} A_i \leq B$.

19. Partition into Triangles [57]

Input: A graph G with 3n nodes.

Property: The nodes of G can be covered with n 3-node complete

subgraphs of G.

20. Rooted Tree Schedule [29]

Input: A rooted tree T, positive integers D and R, and an

assignment to each node i a positive integer r_i.

Property: The nodes of T can be labelled with integers from the set

$\{1,2,\ldots,D\}$ such that no label occurs on more than two nodes,

the sequence of labels or any root-leaf path is strictly

increasing and, for each $j \varepsilon \{1,2,\ldots,D\}$, the sum of the r_i

on nodes labelled j is $\leq R$.

21. Scheduling with Precedence Constraints [60]

Input: A finite partially ordered set S, and positive integers k

and D.

Property: The elements of S can be labelled with integers from the

set $\{1,2,\ldots,D\}$ such that no label occurs more than k times

and, along any ascending chain, the sequence of labels is

strictly increasing.

22. Kernel [6]

Input: A directed graph G.

R. M. Karp

Property: There is a set of nodes K such that no arc joins two elements of K and, for every node u not in K, there is an arc from u to some node in K.

23. Consecutive Sets [46]

Input: A set Σ, a collection $\{S_1, S_2, \ldots, S_n\}$ of finite subsets of Σ, and a positive integer D.

Property: There is a string x of elements of Σ (possibly with repetitions) such that the length of x is D and, for each i, the elements of S_i occur in some order as a block (consecutive subsequence) in x.

24. Discrete Multicommodity Flow [45]

See section entitled Problem Transformations.

25. Steiner Tree [41]

Input: A graph G, a set N of nodes of G, a labelling of the arcs of G with positive integer weights, and a positive integer W.

Property: G contains a tree which includes the set of nodes N, and whose arcs have total weight $\leq$ W.

Prospects

Cook's Theorem and the subsequent work on NP-complete problems raise a number of important theoretical questions. The central question, of course, is whether P = NP. This appears to be a problem that will require the development of entirely new techniques in computational complexity theory. Some useful preliminary work has been done, clarifying the properties of different types of transformations or reductions between problems [46] and giving alternate characterizations of P and NP [7], [9], [23]. A second important objective is to clarify the status of important problems such as the graph isomorphism problem and the linear programming problem by either finding polynomial-time algorithms or proving NP-completeness. Finding new NP-complete problems is a game everyone can play -- it requires no knowledge of computational complexity, but simply the ability to find new polynomial-time problem transformations.

Even if it turns out, as we expect, that $P \neq NP$, the study of combinatorial algorithms for hard problems will still be very much alive. The approach we have outlined deals with the worst-case performance of algorithms as the size of the input goes to infinity. It says nothing about algorithms that solve problems within a specified close approximation, or work most of the time, or work efficiently on inputs of moderate size. For example, it is known that the simplex method for linear programming does not, in its worst case, run in polynomial time [44], [62], but everyone knows that the simplex method can be relied upon to solve efficiently the problems that come up in practice.

There are recent results which show that, for certain hard minimization problems, very fast approximation algorithms exist which are guaranteed to yield solutions within a specified close tolerance of the optimal value. Examples include the "bin packing" [27], travelling salesman [56], knapsack and max-cut [36],[37] problems, which are solvable in polynomial time if and only if $P = NP$, and a two-dimensional placement problem [43] for which no polynomial time exact solution method is known. It would be useful to find further examples of the "fast approximation" phenomenon.

One of the great mysteries in the field of combinatorial algorithms is the baffling success of many heuristic algorithms [50]. The ultimate explanation of this phenomenon will undoubtedly have to be probabilistic. A possible line of approach is suggested by the body of work on random

R. M. Karp

graphs (cf. [20],[21]), which studies the sample space of n-nodes graphs in which each arc is present with probability p, independent of the other arcs. One of the main discoveries of this theory is that, for certain properties such as being connected, containing a clique of a given size, containing a perfect matching, etc., there is a "threshold function" T(n), such that, as n → ∞, the probability of a graph possessing the property converges to 1 if the expected number of arcs is greater than T(n), and to 0 if the expected number of arcs is less than T(n). Perhaps we will be able to find such threshold functions for the probability that simple heuristic algorithms will succeed in finding a clique of given size, a Hamilton circuit, etc., and will be able to gauge the effectiveness of these heuristic algorithms by comparing their thresholds with the known intrinsic thresholds for these problems.

Finally, we must recognize the usefulness of cutting plane methods, branch-and-bound methods and dynamic programming in solving problems of moderate size. There is much that can be done to delay or attenuate the eventual exponential growth in the running times of these methods.

7.   The Fast Approximate Solution of Hard Combinatorial Problems[†]

Introduction

The study of fast algorithms to solve graph-theoretic and combinatorial problems is a rapidly growing field of research. The recent progress is due partly to the exploitation of theoretical results concerning structures such as flow networks, matroids and matchings, and partly to our growing understanding of how to represent data and conduct efficient searches through data structures. In much of this research it has become

[†]Reprinted by permission from Proc. 6th Southeastern Conference on Graph Theory, Combinatorics and Computing, Utilitas Mathematica Publishing Company, Winnipeg (1975).

R. M. Karp

conventional to regard an algorithm as <u>fast</u> if its running time is bounded
by a polynomial in the size of the input. There are fast algorithms, for
example, to find blocks and cut vertices, to test whether a graph is planar, and to compute shortest paths, minimum spanning trees, minimum-cost
flows, and maximum matchings. On the other hand, we do not have fast
algorithms to test the isomorphism of graphs, to solve integer programming
problems or to find maximum cliques, minimum colorings or shortest traveling-salesman tours.

A large class of apparently difficult combinatorial problems have
been proven equivalent, in the sense that either all or none of them can
be solved by fast algorithms. This class includes most of the well-known
combinatorial problems that have proved resistant to efficient computational solution. There is strong theoretical evidence suggesting that
fast algorithms do not exist to solve these problems. In the present
paper we refer to these problems as <u>hard</u>. Hard problems are usually
attacked in practice through so-called heuristic algorithms. These
algorithms are fast, and often yield near-optimal solutions, but lack a
theoretical foundation.

There is recent work suggesting that a rigorous theory of heuristic
algorithms may be attainable. Such a theory will be concerned with fast
algorithms which solve a problem within a specified approximation, or with
a specified probability. Thus, given a minimization problem and a real
number $r \geq 1$, an algorithm is said to <u>solve the problem within</u> r if it
produces a feasible solution whose cost is not more than r times the
cost of an optimal feasible solution. Given a probability distribution
over the possible inputs, we may seek algorithms that "almost surely"
solve the problem within r. This paper surveys recent work concerning

the existence of fast algorithms for solving hard combinatorial problems
according to these criteria.

The next section gives a brisk survey of some of the principal fast
graph-theoretic algorithms. It then introduces the concepts of polynomial-
time reducibility and NP-completeness, and reviews the evidence indicating
that NP-complete problems cannot be solved by fast algorithms. The subsection entitled <u>Approximate Optimization</u> is concerned with approximate
algorithms. Fast algorithms are given for solving the subset-sum problem
within $1 + e$, the bin packing problem within $\frac{11}{9}$, and the metric traveling-salesman problem within 2. A negative result is also given, indicating that a fast algorithm to solve graph coloring problems within $r < 2$
is unlikely to exist. The subsection entitled <u>Probabilistic Models</u> deals
with probabilistic models. Fast algorithms are given which almost surely
solve the euclidean traveling-salesman problem within $1 + e$ and the graph-
coloring problem within $2 + e$. The probabilistic behavior of an algorithm
for finding Hamilton circuits is also discussed.

Fast Algorithms and NP-Complete Problems

In a basic paper on matching theory [16] Jack Edmonds first raised
the question of the existence of polynomial-time algorithms (he called
them "good" algorithms) for solving problems in graph theory. In other
contexts the distinction between polynomial-time and non-polynomial-time
algorithms was also stressed by Cobham [7] and Rabin. We shall refer to
polynomial-time bounded algorithms as <u>fast</u>.

We shall indicate how the concept of a fast algorithm is formalized.
We suppose that each possible specification of the input data for an
algorithm is encoded as a string of zeros and ones (any other fixed finite
alphabet of two or more symbols would serve equally well). Thus an integer

might be represented in binary and a graph by a string of symbols giving its adjacency matrix or a list of its edges. Algorithms are defined using a simplified abstract model of a digital computer, called a RAM (random access machine) [10]. An algorithm is <u>fast</u> if there is a polynomial $f(\cdot)$ such that, on any input of length n, the algorithm executes at most $f(n)$ steps before halting.

Network flow theory is the most fruitful source of fast algorithms in graph theory. It yields fast algorithms for finding shortest paths in graphs, and for solving assignment, transportation and minimum-cost flow problems. During the period 1953-1962, when network flow theory underwent its most active development, it was not customary to analyze the worst-case running time of an algorithm. Algorithms were compared empirically or on aesthetic grounds. In fact, most of the original network flow algorithms did not run in polynomial time, although many of them were later modified to do so ([14],[19],[22]). Matching theory and matroid theory also proved to be fruitful sources of fast algorithms; algorithms for maximum matchings, maximum matroid intersections and minimum matroid partitions are among the most important of these algorithms ([16],[17], [49]).

Starting around 1970 an important line of research developed, aimed at making fast algorithms even faster, principally through the clever use of data structures. Thus, given a graph with v vertices and e edges, it is easy to construct a minimum spanning tree within $O(e \log v)$ steps, but altering the algorithm to obtain a time bound of $O(e \log \log v)$ requires considerably subtlety ([59],[61]). Edmonds' original algorithm for computing maximum cardinality matchings requires $O(v^4)$ steps, but a different implementation reduces the time to $O(v^{5/2})$ ([22]). Through

R. M. Karp

the use of appropriate data structures and search techniques, algorithms
running in time $O(v+e)$ were obtained for finding the strong components
of a digraph, and for testing whether a graph is planar [34], [4], chor-
dal [55], an interval graph [4], doubly connected [58] or triply connected [33]. Corneil [11] gives a good survey of fast graph-theoretic
algorithms.

Many basic algorithmic problems have resisted the search for fast
algorithms. These include the traveling-salesman problem, the deter-
mination of maximum cliques and minimum vertex colorings in graphs, the
graph isomorphism problem, and various other problems of packing, covering, routing and sequencing. It has been shown ([41],[30],[42]) that a
very large number of these problems are equivalent, in the sense that
either all or none of them can be solved in polynomial time. Moreover,
there is theoretical evidence which makes it unlikely that any of these
problems can be solved in polynomial time. My account of these develop-
ments will be brief, since they have been surveyed elsewhere [42].

We distinguish recognition algorithms, which terminate by signalling
whether an input string is accepted or rejected, from algorithms which
map input strings onto output strings. Thus an algorithm to decide whether a graph is Hamiltonian is a recognition algorithm, whereas an
algorithm to compute a minimum-cost traveling-salesman tour is not. Let
$\{0,1\}^*$ denote the set of all finite strings of 0's and 1's. With any
recognition algorithm we may associate a set $L \subseteq \{0,1\}^*$ consisting of
those input strings accepted by the algorithm; L is called the language
defined by the recognition algorithm. Similarly, any string-mapping
algorithm defines a function from $\{0,1\}^*$ into $\{0,1\}^*$. We define P
as the family of all languages defined by fast recognition algorithms,

R. M. Karp

and Ⅱ as the set of all functions defined by fast string-mapping algorithms.

We are now ready to define <u>polynomial-time reducibility</u>, a concept which permits us to partially order languages according to their computational complexity. Given two sets $L \subseteq \{0,1\}^*$ and $M \subseteq \{0,1\}^*$, we say $L \propto M$ (L <u>is reducible to</u> M) if there is a function $f \in \Pi$ such that, for all $x \in \{0,1\}^*$, $x \in L \Leftrightarrow f(x) \in M$. If $L \propto M$ then we can decide if $x \in L$ by computing $f(x)$, and then testing whether $f(x) \in M$. It follows that, if $L \propto M$ and $M \in P$, then $L \in P$; i.e., if a problem is easily converted into an easy problem, then it is also an easy problem. Define an equivalence relation $\equiv$ by: $L \equiv M$ if $L \propto M$ and $M \propto L$. Then all the elements of an equivalence class are in P if any particular member of the class is in P.

Most of the apparently difficult recognition problems of importance in graph theory and combinatorics lie in a single equivalence class. Typical members of this class are the Hamilton circuit problem, the problem of deciding whether a graph has a clique of a given size or a coloring with a given number of colors, and the problem of deciding whether a system of linear inequalities has a solution in integers. The members of this class are called <u>NP-complete problems</u>.

Cook [8] has proved an important theorem which suggests that the NP-complete problems are unlikely to be in P. He defines NP as the class of recognition problems solvable in polynomial time by nondeterministic algorithms, and shows that every problem in NP is reducible to every NP-complete problem. The precise definition of "nondeterministic algorithm" would take us far afield; informally, a nondeterministic algorithm is one which can generate all the paths through a backtrack

R. M. Karp

search tree in parallel, accepting the input if any path leads to success.
Thus a decision problem is in NP if it can be solved by a backtrack
search of polynomial bounded depth. It follows that the NP-complete problems are in P if and only if $P = NP$.

Nearly every recognition problem arising in the applications of
graph theory is in NP. Since so many of these computational problems
have been resistant to solution by fast algorithms, it seems likely that
$P \neq NP$; i.e., that the NP-complete problems cannot be solved by fast
algorithms.

Let us call a problem hard if the existence of a fast algorithm for
its solution implies $P = NP$. The NP-complete problems are hard, of
course. Many optimization problems are also hard. An example is the
problem of computing a tour of minimum length, given the distances
between all pairs of vertices. If there were a fast algorithm for this
problem, it could be used to decide quickly whether a graph has a Hamilton
circuit. Since the latter problem is NP-complete, it would then follow
that $P = NP$.

Approximate Optimization

This section is concerned with fast algorithms for the approximate
solution of hard optimization problems.

Solving the Subset-Sum Problem Within 1+ε

An instance of the subset-sum problem is specified by a n-tuple of
positive rational numbers $\underline{a} = (a_1, a_2, \ldots, a_n)$ and a positive rational
number b. The problem is to determine a set $T \subseteq \{1, 2, \ldots, n\}$ such that
$\sum_{i \in T} a_i$ is maximized, subject to the constraint $\sum_{i \in T} a_i \leq b$. It is known that
the problem of deciding whether there is a T such that $\sum_{i \in T} a_i = b$ is

R. M. Karp

NP-complete. Hence the subset-sum problem is hard.

As a first step toward a fast approximate optimization algorithm, we consider the following related decision problem: given $\underline{a} = (a_1, a_2, \ldots, a_n)$ a positive rational number c and a "tolerance" Δ, determine whether there exists a set T such that $c - \Delta \leq \sum_{i \in T} a_i \leq c$. We give an algorithm for solving this problem in about $\frac{nc}{\Delta}$ steps.

Given any set $S \subseteq [0,c]$, and any rational d, let $S + d = \{x+d \mid x \in S \text{ and } x+d \in [0,c]\}$. Let $N(S)$, the Δ-neighborhood of S, be given by

$$N(S) = \{x \mid x \in [0,c] \text{ and, for some } y \in S, \; |x-y| \leq \Delta\} .$$

For any i, $i = 1,2,\ldots,n$, let $R(i)$, the ith reachability set, be defined as

$$R(i) = \{x \mid x \in [0,c] \text{ and, for some } U \subseteq \{1,2,\ldots,i\}, \; \sum_{i \in U} a_i = x\} .$$

Then our problem is to decide whether $c \in N(R(n))$.

The set $R(n)$ is in general rather complicated, but for purposes of determining $N(R(n))$, we may replace it with a set $\overline{R(n)}$ which is much easier to describe. For any set $S \subseteq [0,c]$, define $\overline{S}$, the Δ-closure of S, by

$$\overline{S} = \{x \in [0,c] \mid \text{ for some } y \in S \text{ and } z \in S, \; y \leq x \leq z \leq y+2\Delta\} .$$

Then $\overline{S}$ is the union of at most $\frac{c}{2\Delta}+1$ closed intervals, each of which can be specified by its end points. Also, $N(\overline{S}) = N(S)$. Hence we may proceed by computing $\overline{R(1)}, \overline{R(2)}, \ldots, \overline{R(n)}$, and then $N(\overline{R(n)})$. We obtain $\overline{R(i+1)}$ from $\overline{R(i)}$ using the relation

$$\overline{R(i+1)} = \overline{\overline{R(i)} \cup (\overline{R(i)} + a_{i+1})} .$$

R. M. Karp

This entire computation can be performed in $O(\frac{nc}{\Delta})$ steps, and requires about $\frac{2c}{\Delta}$ cells of temporary storage.

Example.

$\quad$ c = 90 $\qquad \Delta$ = 6 $\qquad$ n = 4 $\qquad (a_1,\ldots,a_4)$ = (17,20,32,33)

$\quad$ R(4) = {0,17,20,32,33,37,49,50,52,53,65,69,70,82,85}

$\quad \overline{R(3)}$ = {0} $\cup$ [17-52] $\cup$ {69}

$\quad \overline{R(4)}$ = ({0}$\cup$[17-52]$\cup${69}) $\cup$ ({33}$\cup$[50-85]) = {0}$\cup$[17,85]

$\quad N(\overline{R(4)})$ = N(R(4)) = [0,6]$\cup$[11,90]

Using this decision method we solve the subset-sum problem within $1 + \epsilon$ as follows. If $\sum\limits_{i=1}^{n} a_i \leq b$ then $\sum\limits_{i=1}^{n} a_i$ is the optimal value for the subset-sum problem. Otherwise the optimal value lies between $\frac{b}{2}$ and b. Using the decision method, find the least k for which there is a T such that $b(1+\epsilon)^{-(k+1)} \leq \sum\limits_{i\in T} a_i \leq b(1+\epsilon)^{-k}$; to make this test for a given k, set $c = b(1+\epsilon)^{-k}$ and $\Delta = b(1+\epsilon)^{-(k-1)}$. The running time for each application of the decision procedure is $O(\frac{n(1+\epsilon)}{\epsilon})$, and the number of unsuccessful applications of the procedure cannot exceed $\log_{1+\epsilon} 2$. Hence the running time of the algorithm for solving the subset-sum problem within $1 + \epsilon$ is about $n(\frac{1+\epsilon}{2})\log_{1+\epsilon} 2$. The temporary storage requirement is about $2(\frac{1+\epsilon}{\epsilon})$ cells.

Previous work on approximate solutions to the subset-sum problem can be found in [38] and [36]. The subset-sum problem is a special case of the knapsack problem, which can also be solved within $1 + \epsilon$ by a fast algorithm [36].

Solving the Bin Packing Problem Within 11/9

Suppose we are given n objects to be packed into bins. The weight of each object is given, and no more than W units of weight may be

R. M. Karp

packed into any given bin. What is the minimum number of bins necessary? For example, suppose W = 4 and we have objects of the following types: 6 objects of weight 2.1, 6 objects of weight 1.2, 6 objects of weight 1.1 and 12 objects of weight .8. The following figure exhibits a packing using 9 bins. Moreover. this packing cannot be improved, since it fills each of the 9 bins completely.

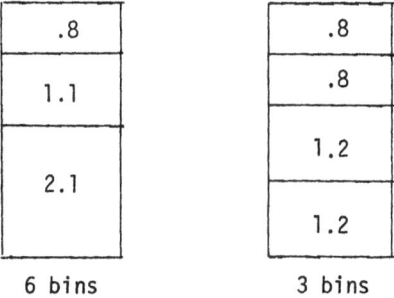

Fig. 1. A Perfect Packing Using 9 Bins

Since even the simple question of determining whether 2 bins suffice is NP-complete, the bin packing problem is hard.

Consider the following simple "First-Fit Decreasing" packing procedure: arrange the objects in order of decreasing size, and pack each object in turn into the first bin that will accept it. Applying this heuristic to the above example, we obtain the following packing.

R. M. Karp

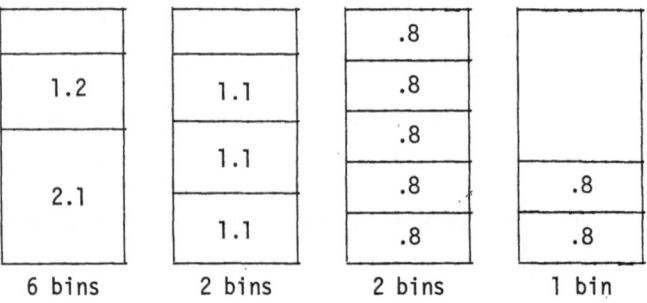

Fig. 2.   A Packing Using 11 Bins

Let N_{OPT} denote the number of bins used in an optimal solution, and let N_{FFD} denote the number used by the first-fit decreasing procedure. Then $N_{FFD} = 11$, $N_{OPT} = 9$ and $\frac{N_{FFD}}{N_{OPT}} = \frac{11}{9}$. Surprisingly, the ratio $\frac{11}{9}$ represents the worst possible behavior of the first-fit decreasing procedure. A theorem due to David Johnson [39] shows that, in all cases $N_{FFD} \leq \frac{11}{9} N_{OPT} + 4$.

Solving the Metric Traveling-Salesman Problem Within 2

Given n points, together with the distance between each pair of points, the traveling-salesman problem asks for a simple cycle of minimum total distance passing through all n points. We consider the case where the distances have the properties of a metric; if d_{ij} denotes the distance between i and j, we have:

$$d_{ii} = 0 , \quad d_{ij} = d_{ji} , \quad d_{ij} + d_{jk} \geq d_{ik} .$$

It is easy to prove that the metric traveling-salesman problem is hard. To solve it within 2, simply:

(i)   construct a minimum-weight spanning tree T through 1,2,...,n, using d_{ij} as the weight of edge {i,j};

(ii)   construct a closed walk W passing along each edge of T exactly twice;

(iii) convert W to a simple cycle C by repeated application of the following "bypass" operation: replace the two edges {i,k} and {k,j} by a single edge {i,j}.

Figure 3 illustrates steps (i) and (ii) of the construction for a case where the cities are points in the plane, and the d_{ij} represent euclidean distances.

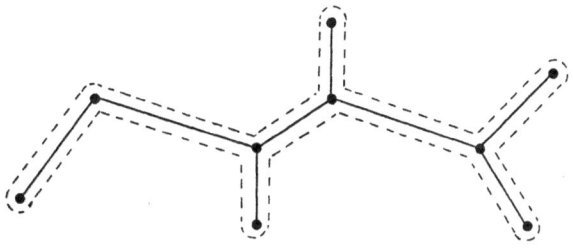

Fig. 3.   Closed Walk Constructed From a Minimum Spanning Tree

The weight of W is twice the weight of T; and the triangle inequality implies that the weight of C is less than or equal to the weight of W. On the other hand every simple cycle through all the vertices contains a spanning tree, and thus has weight greater than or equal to the weight of T. It follows that the cost of an optimal solution is at least the weight of T. Hence C solves the metric traveling-salesman problem within 2. The construction we have given has been noted by Eugene Lawler, Nicholas Pippinger and others. Other efficient methods of solving the metric traveling-salesman problem within 2 are given in [56]. It is an open question whether, for any r < 2, there exists a fast algorithm which solves the metric traveling-salesman problem within r. Also open is the question of fast methods for approximate solution of the directed traveling-salesman problem in which the distances satisfy the triangle inequality.

A Negative Result About Graph Coloring

A striking negative result about the difficult of approximately
optimal graph coloring has been obtained by Garey and Johnson [28].
Let $\chi(G)$ denote the chromatic number of the graph G. They showed
that if, for some $r < 2$ and $d > 0$, there exists a fast algorithm
which colors any graph G with at most $r\chi(G) + d$ colors, then
$P = NP$. This result suggests that solving the graph coloring problem
within $r < 2$ is as hard as solving it exactly.

Probabilistic Models of Approximate Optmization Methods

Measured by the criteria normally used in computational complexity
theory, the simplex method of solving linear programs has little merit.
Klee and Minty [44], Jeroslow and Zadeh [62] have constructed families
of examples in which the simplex method inspects all, or nearly all,
the vertices of a polyhedron in order to find the vertex at which a
given linear function is maximized. Nevertheless the simplex method is
the backbone of computational practice in mathematical programming, and
is observed to perform with unfailing efficiency on the real-world
computational problems presented to it. Similarly, users of heuristic
algorithms for program solving in combinatorics or in artificial intelli-
gence are troubled little by the fact that these algorithms can fail
badly on selected inputs, so long as these inputs occur rarely in
practice.

In this section we give three examples of the probabilistic analysis of fast algorithms for solving hard combinatorial problems. In
each case we assume that the input data presented to the algorithm is

drawn from a probability distribution, and we conclude that the algorithm "almost surely" is successful. While recognizing that it is guesswork to assume any particular probability distribution of problem data, we feel that a probabilistic approach can provide insights into the workings of heuristic algorithms that are not accessible through the traditional method of worst-case analysis.

Methodology

The ingredients of a probabilistic model of an algorithm are:

(i)   the algorithm;

(ii) a criterion of successful performance, such as solving the problem exactly, or within r

and

(iii) a sample space of inputs.

In setting up the sample space, we must take into account the fact that the algorithm can accept inputs of unbounded size -- graphs with any number of vertices, for example. We can avoid making specific assumptions about the distribution of problem sizes by setting up the sample space in two stages, as follows. First, for each n, we define a sample space S_n from which inputs of size n are drawn. Then our overall sample space is the infinite cartesian product $S_1 \times S_2 \times \cdots \times S_n \times \cdots$. Thus an element of the sample space is a sequence $X = x_1, x_2, \ldots, x_n, \ldots$, where x_n is drawn from S_n. We say that the algorithm succeeds almost surely if, when $X = x_1, x_2, \ldots, x_n, \ldots$ is drawn from the cartesian product distribution, the number of x_n on which the algorithm fails to meet its performance criterion is finite with probability 1.

R. M. Karp

The Euclidean Traveling-Salesman Problem

Garey and Johnson have proved that finding a tour of minimum euclidean length through a set of n points in the plane is hard. Since
the euclidean traveling-salesman problem is a special case of the metric
traveling-salesman problem, there is a fast algorithm to solve it within
2. Here we exhibit a fast algorithm which, for every e > 0, almost
surely solves the euclidean traveling-salesman problem within 1 + e.

We assume that a random problem of size n is specified by n
points drawn at random from the unit square. Our algorithm is based on
a divide-and-conquer principle. It depends on the fact that there is a
dynamic programming algorithm for solving an arbitrary t-city traveling-
salesman problem within $O(t^2 \cdot 2^t)$ steps [2]. Let $\lambda(n)$ be an unbounded
nondecreasing positive function. Given n, choose the least k such
that $\frac{n}{k^2} \leq \lambda(n)$. Subdivide the unit square into a checkerboard consisting of k^2 smaller squares. Using the dynamic programming algorithm,
find a minimum-length polygon through the set of points that fall into
each square of the checkerboard. Then, using the minimum spanning tree
algorithm, find a set of line segments of minimum total length that
join the k^2 polygons together into a connected figure. The original
tours, plus two copies of each of the line segments in the minimum connecting set, determine a closed walk through all the points. By repeated
use of the bypass operation discussed in the section entitled Approximate Optimization, convert this closed walk into a tour.

A detailed analysis of the performance of this algorithm will not
be presented here. The analysis depends on theorems by Beardwood,
Halton and Hammersley [1] concerning the probability distribution of
the length of a shortest tour through n random points. The typical

running time of the algorithm is $O(ne^{\lambda(n)}\lambda(n))$, and it typically

solves the problem within $1 + \dfrac{constant}{\lambda(n)^{5/6}}$. Choosing $\lambda(n) = \log n$, we

obtain a fast algorithm that, for every ϵ, almost surely solves the

euclidean traveling-salesman problem within $1 + \epsilon$.

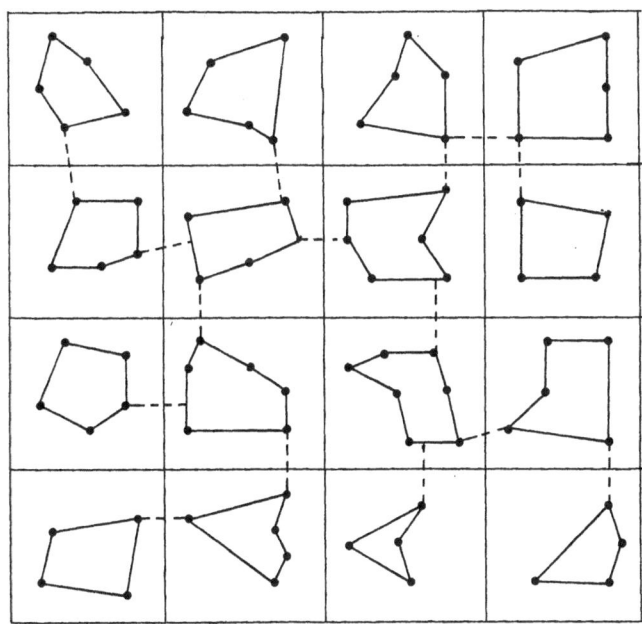

Fig. 4.   Polygons Plus Two Copies of Dotted
Lines Constitute a Closed Walk

Cliques and Colorings

The problems of finding a clique of maximum size or a coloring

with a minimum number of colors in a graph are both hard.   The theorem

of Garey and Johnson in the section entitled <u>Approximate Optimizations</u>

shows that solving the coloring problem within $r < 2$ is also hard.

Johnson [40] has surveyed several fast heuristic methods of finding

large cliques and economical colorings, and has noted that their worst-

case performance is extremely bad.   Here we introduce two of these methods, and state results due to Grimmett and McDiarmid [31] showing that they almost surely perform reasonably well.

The <u>sequential method</u> of finding a large clique examines the vertices in random order, and adds each in turn to the clique being constructed provided it is adjacent to all the vertices already in the clique. On a graph such as the one in Fig. 5, the algorithm will do very poorly if u and v happen to be the first two vertices considered.

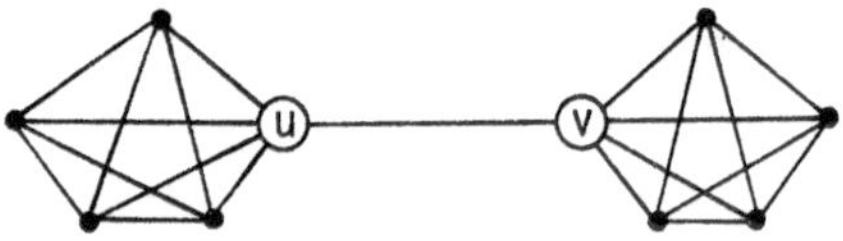

Fig. 5. A Bad Example for the Sequential
Clique-Finding Method

We outline a proof that the sequential method almost surely solves the maximum clique problem within 2. This proof, which differs slightly from the one by Grimmett and McDiarmid, is due to Matula and the writer. We assume that a random n-vertex graph is generated as follows: for each pair of vertices $\{i,j\}$, the event that $\{i,j\}$ is an edge of G has probability p; the $\binom{n}{2}$ events corresponding to all the pairs $\{i,j\}$ are independent. The probability p is fixed, independent of n. Matula [51] has shown that the size of the largest clique in a random graph has a highly spiked distribution -- it is almost surely within 1 of the quantity

$$2 \log_{1/p} n - 2 \log_{1/p} \log_{1/p} n + 2 \log_{1/p} (\tfrac{e}{2}) + 1 \ .$$

On the other hand, the sequential method almost surely yields a clique of size at least

$$\log_{1/p} n - \log_{1/p} \log_{1/p} n \ .$$

The result follows by combining these two observations.

R. M. Karp

A similar approach can be taken to graph coloring. Let us name the colors by the successive positive integers $1,2,\dots$. The sequential coloring method considers the vertices in random order, and assigns to each vertex in turn the least color which has not been assigned to any vertex adjacent to it. While the sequential method performs very poorly on perversely chosen examples, one can show that it almost surely solves the coloring problem within $2+e$, for every $e > 0$.

Hamilton Circuits

Let α be a positive constant. Let a random n-vertex graph be constructed by selecting $[\alpha n \ln n]$ edges out of the $\binom{n}{2}$ possible edges. A well-known result by Erdös and Renyi [20] asserts that, if $\alpha < \frac{1}{2}$, the graph almost surely is not connected; hence it almost surely has no Hamilton circuit. Posá [52], Komlos and Szemeredi have recently shown that, if $\alpha > \frac{1}{2}$, the graph almost surely does have a Hamilton circuit. The result is of interest in the present context because of its proof. The proof consists, in essence, of the probabilistic analysis of an algorithm for constructing Hamilton circuits. The fact that the algorithm is fast was not pointed out in [52], since it was not required for the proof given there.

Given an n-vertex graph G, the algorithm attempts to construct longer and longer simple paths starting at vertex 1. In doing so it uses the operations of _extension_ and _rotation_. Let P be a path joining vertices 1 and v; v is called the _free end point_ of P. Given an edge {v,w}, $w \neq 1$, which is not in P, we distinguish two cases:

(i) if w does not occur in P, then the operation of extension applies; this operation simply adjoins {v,w} to P, producing a path from 1 to w;

(ii) if w occurs in P, then the rotation operation adjoins {v,w} to P, and deletes that unique edge {w,x} in P such that P∪{v,w} - {w,x} is a path.

Having found a path of length i, the algorithm creates a sequence of new paths of length i by rotation, continuing until one of the paths admits of extension. The process continues until a Hamilton line has been found; it then generates further Hamilton lines by rotation, seeking to extend some such line to a circuit.

A restriction is imposed on the types of rotations permitted. Suppose the initial path of length i is P_0. As the process of rotation proceeds, a list L is maintained which includes the free end points of the successive paths generated, along with the neighbors of those end points in P_0. No rotation is permitted which would connect the free end point of some existing path to a vertex in L. This restriction implies that no two paths of length i have the same free end point. Hence not more than n-1 paths of length i are generated; a polynomial time bound for the algorithm follows from this fact.

Inspection of the algorithm shows that it cannot fail to find a Hamilton circuit unless the graph possesses two disjoint sets of vertices, A and B, such that

(i) for some i between 1 and n-1,

$$|B| \geq n - i + \max(0, i-3|A|)$$

and
(ii) no edge joins a vertex in A with a vertex in B.
Such a pair of sets almost surely does not exist if $\alpha > \frac{1}{2}$. The algorithm has been tested by Robert MacGregor and the writer. These tests confirm that the algorithm is an effective method for finding Hamilton circuits in random graphs.

R. M. Karp

1. J. Beardwood, J.H. Halton and J.M. Hammersley, "The Shortest Path Through Many Points," *Proc. Camb. Phil. Soc.* 55 (1959), 299-327.

2. R. Bellman, "Combinatorial Processes and Dynamic Programming," *Proc. Tenth Symp. in Applied Math.*, American Mathematical Society (1960), 217-250.

3. C. Berge, "Two Theorems in Graph Theory," *Proc. Nat. Acad. Sci.* 43 (1957), 842-844.

4. K.S. Booth and G.S. Lueker, "PQ-Tree Algorithms," submitted to *J. Comp. and Syst. Sci.* (1975).

5. G.H. Bradley, "Equivalent Integer Programs and Canonical Problems," *Man. Sci.* 17 (1971), 354-366.

6. V. Chvátal, personal communication, 1973.

7. A. Cobham, "The Intrinsic Computational Difficulty of Functions," *Logic, Methodology and Philosophy of Science*, North-Holland (1965).

8. S.A. Cook, "The Complexity of Theorem-Proving Procedures," *Proc. Third ACM Symp. on Theory of Computing* (1971), 151-158.

9. S.A. Cook, "Characterizations of Pushdown Machines in Terms of Time-Bounded Computers," *J. Assoc. Comp. Mach.* 18 (1971), 4-18.

10. S.A. Cook and R.A. Reckhow, "Time Bounded Random Access Machines," *J. Comp. and Syst. Sci.* 7 (1973), 354-375.

11. D. Corneil, see *Proc. Fifth Southeastern Conference on Combinatorics, Graph Theory and Computing*, Utilitas Mathematica Publishing, Winnipeg (1974).

12. G.B. Dantzig, "On the Significance of Solving Linear Programming Problems with Some Integer Variables," *Econometrica* 28 (1960), 30-44.

13. E. Dijkstra, "Note on Two Problems in Connection with Graphs," *Numer. Math.* 1 (1959), 269-271.

14. E.A. Dinic, "Algorithm for Solution of a Problem of Maximum Flow in a Network with Power Estimation," *Sov. Math. Doklad.* 11 (1970), 1277-1280.

15. S.E. Dreyfus, "An Appraisal of Some Shortest Path Algorithms," *Operations Research* 17 (1969), 395-412.

16. J. Edmonds, "Paths, Trees and Flowers," *Canad. J. Math.* 17 (1965), 449-467.

17. J. Edmonds, "Minimum Partition of a Matroid into Independent Subsets, *J. Res. NBS* 69B (1965), 67-72.

18. J. Edmonds and E.L. Johnson, "Matching: A Well-Solved Class of Integer Linear Programs," *Combinatorial Structures and Their Applications*, Gordon and Breach (1970), 89-92.

19. J. Edmonds and R.M. Karp, "Theoretical Improvements in Algorithmic Efficiency for Network Flow Problems," *J. Assoc. Comp. Mach.* 9 (1972), 248-264.

20. P. Erdös and A. Renyi, "On Random Graphs, I," *Publicationes Mathematicae* 6 (1959), 290-297.

21. P. Erdös and J. Spencer, *Probabilistic Methods in Combinatorics*, Academic Press (1974).

22. S. Even and O. Kariv, "An $O(n^{5/2})$ Algorithm for Maximum Matchings in General Graphs," *Proc. Sixteenth Annual Symp. on Foundations of Computer Science, IEEE* (1975), 100-112.

23. R. Fagin, "Generalized First-Order Spectra and Polynomial-Time Recognizable Sets," *Complexity of Computation*, ed. R. Karp, *Proc. SIAM-AMS Symposia in Applied Mathematics* (1974).

24. R.W. Floyd, "Nondeterministic Algorithms," *J. Assoc. Comp. Mach.* 14 (1967), 636-644.

25. L.R. Ford and D.R. Fulkerson, *Flows in Networks*, Princeton University Press (1962).

26. H. Gabow, "An Efficient Implementation of Edmonds' Maximum Matching Algorithm," *Technical Report No. 31*, Stanford Digital Systems Laboratory (1972).

27. M.R. Garey, R.L. Graham and J.D. Ullman, "Worst-Case Analysis of Memory Allocation Algorithms," *Proc. Fourth ACM Symposium on Theory of Computing* (1972), 143-150.

28. M.R. Garey and D.S. Johnson, "The Efficiency of Near-Optimal Graph Coloring," Bell Laboratories Report (1974).

29. M.R. Garey and D.S. Johnson, "Complexity Results for Multiprocessor Scheduling Under Resource Constraints," Bell Telephone Laboratories (1974).

30. M.R. Garey, D.S. Johnson and L. Stockmeyer, "Some Simplified NP-Complete Graph Problems," *Proc. Sixth ACM Symposium on Theory of Computing* (1974), 47-63; to appear in *Theoretical Computer Science*.

31. G.R. Grimmett and C.J.H. McDiarmid, "On Colouring Random Graphs," *Math. Proc. Camb. Phil. Soc.* 77 (1975), 313-324.

32. J.E. Hopcroft and R.M. Karp, "An $n^{5/2}$ Algorithm for Maximum Matchings in Bipartite Graphs," *SIAM J. Comput.* 2 (1973), 225-231.

33. J.E. Hopcroft and R.E. Tarjan, "Dividing a Graph Into Triconnected Components," *SIAM J. Comput.* 2 (1973), 135-158.

34. J.E. Hopcroft and R.E. Tarjan, "Efficient Planarity Testing," *J. Assoc. Comput. Mach.* 21 (1974), 549-568.

35. J.E. Hopcroft and J.D. Ullman, *Formal Languages and Their Relation to Automata*, Addison-Wesley, Reading, Mass. (1969).

36. O.H. Ibarra and C.E. Kim, "Fast Approximation Algorithms for the Knapsack and Sum of Subset Problems," *J. Assoc. Comput. Mach.* 22 (1975), 463-468.

37. O.H. Ibarra and S. Sahni, "P-Complete Problems and Approximate Solutions," *Technical Report 74-5*, University of Minnesota (1974).

38. D.S. Johnson, "Approximation Algorithms for Combinatorial Problems," to appear in *J. Comp. and Syst. Sci.*

39. D.S. Johnson, "Fast Algorithms for Bin Packing," *J. Comp. and Syst. Sci.* 8 (1974), 272-314.

40. D.S. Johnson, "Worst-Case Behavior of Graph Coloring Algorithms," *Proc. Fifth Southeastern Conference on Combinatorics, Graph Theory and Computing*, Utilitas Mathematica Publishing, Winnipeg (1974).

41. R.M. Karp, "Reducibility Among Combinatorial Problems," *Complexity of Computer Computations*, R.E. Miller and J.W. Thatcher, eds., Plenum Press, New York (1972), 85-104.

42. R.M. Karp, "On the Computational Complexity of Combinatorial Problems," *Networks* 5 (1975), 45-68.

43. R.M. Karp, A.C. McKellar and C.K. Wong, "Near-Optimal Solutions to a 2-Dimensional Placement Problem," *SIAM J. Comput.*, to appear in 1975.

44. V. Klee and G.J. Minty, "How Good is the Simplex Algorithm?," *Mathematical Note No. 643*, Boeing Scientific Research Laboratories (1970).

45. D.E. Knuth, personal communication, 1974.

46. L. Kou, "Polynomial Complete Consecutive Information Retrieval Problems," to appear in *SIAM J. Comput.*

47. R. Ladner, N. Lynch and A.L. Selman, "Comparison of Polynomial-Time Reducibilities," *Proc. Sixth ACM Symposium on Theory of Computing* (1974), 110-121.

48. E.L. Lawler, *Combinatorial Optimization: Networks and Matroids*, Holt, Rinehart and Winston, Inc., to appear.

49. E.L. Lawler, "Matroid Intersection Algorithms," *Mathematical Programming* 9 (1975), 31-56.

50. S. Lin, "Heuristic Programming as an Aid to Network Design," *Proc. Symposium on Large-Scale Networks, Networks* 5 (1975), 33-43.

51. D. Matula, "The Employee Party Problem," abstract, *Amer. Math. Soc. Notices* (1972).

52. L. Posá, "Hamilton Circuits in Random Graphs," to appear in *Discrete Mathematics* (1975).

53. R. Reiter, personal communication, 1971.

54. R. Rivest, personal communication, 1974.

55. D.J. Rose and R.E. Tarjan, "Algorithmic Aspects of Vertex Elimination," *Proc. Seventh Annual ACM Symposium on Theory of Computing* (1975), 245-254.

56. D.J. Rosenkrantz, R.E. Stearns and P.M. Lewis, "Approximation Algorithms for the Traveling-Salesperson Problem," *Proc. Fifteenth IEEE Switching and Automata Theory Symposium* (1974).

57. T. Schaefer, personal communication, 1974.

58. R. Tarjan, "Depth-First Search and Linear Graph Algorithms," *SIAM J. Comput.* 1 (1972), 146-159.

59. R.E. Tarjan, "Finding Minimum Spanning Trees," *ERL-M501*, Electronics Research Laboratory, University of California, Berkeley (1975).

60. J.E. Ullman, "Polynomial Complete Scheduling Problems," *Fourth Symp. on Operating System Principles* (1973), 96-101.

61. A. Yao, "An $O(|E|\log\log|V|)$ Algorithm for Finding Minimum Spanning Trees," to appear in *Inf. Proc. Letters* (1975).

62. N. Zadeh, "A Bad Network Problem for the Simplex Method and Other Minimum Cost Flow Algorithms," *Mathematical Programming* 5 (1973), 255-266.

Editoriale Grafica · Roma

CENTRO INTERNAZIONALE MATEMATICO ESTIVO
(C.I.M.E.)
INTERNATIONAL MATHEMATICAL SUMMER CENTER

VOLUMES ON C.I.M.E. SESSIONS

1968

1. **Controllability and Observability.** Directed by Prof. G. Evangelisti . . . $ 8.—
2. **Pseudo-differential Operators.** Directed by Prof. L. Nirenberg . . . . » 10.—
3. **Aspects of Mathematical Logic.** Directed by Prof. E. Casari . . . . » 8.—

1969

1. **Potential Theory.** Directed by Prof. M. Brelot . . . . . . » 8.—
2. **Non-linear Continuum Theories in Mechanics and Physics and their applications.** Directed by Prof. R. S. Rivlin . . . » 10.—
3. **Questions on Algebraic Varieties.** Directed by Prof. E. Marchionna . . . . » 10.—

1970

1. **Relativistic Fluid Dynamics.** Directed by Prof. C. Cattaneo . . . . » 13.—
2. **Theory of Group Representations and Fourier Analysis.** Directed by Prof. F. Gherardelli . . . . . . » 10.—
3. **Functional Equations and Inequalities.** Directed by Prof. B. Forte. . . . » 13.—
4. **Problems in non-linear Analysis.** Directed by Prof. G. Prodi . . . . » 16.—

1971

1. **Stereodynamics.** Directed by Prof. G. Grioli . . . . » 10.—
2. **Constructive Aspects of Functional Analysis.** Directed by Prof. G. Geymonat. (Being printed) . . . . . » 25.—
3. **Categories and Commutative Algebra.** Directed by Prof. P. Salmon . . . » 10.—

1972

1. **Non-linear Mechanics.** Directed by Prof. D. Graffi . . . . . . » 14.—
2. **Finite geometric structures and their applications.** Directed by Prof. A. Barlotti » 9.—
3. **Geometric measure theory and minimal surfaces.** Directed by Prof. G. Bombieri. » 7.—

1973

1. **Complex Analysis.** Directed by Prof. F. Gherardelli . . . . . . » 14.—
2. **New variational techniques in mathematical physics.** Directed by Prof. G. Capriz and Prof. G. Stampacchia . . . » 10.—
3. **Spectral Analysis.** Directed by Prof. J. Cecconi . . . . » 9.—

1974

1. **Stability problems.** Directed by prof. L. Salvadori . . . . . . » 9.—
2. **Singularities of analytic spaces.** Directed by prof. A. Tognoli . . . . » 8.—
3. **Eigenvalues of non-linear problems.** Directed by prof. G. Prodi . . . . » 9.—

EDIZIONI CREMONESE
Via della Croce, 77
00187 ROMA (Italia)